THE TECHNOLOGY TAKERS

THE TECHNOLOGY TAKERS

Leading Change in the Digital Era

BY

JENS P. FLANDING, Ph.D.
GENEVIEVE M. GRABMAN, J.D.
SHEILA Q. COX, MBA

United Kingdom — North America — Japan
India — Malaysia — China

Emerald Publishing Limited
Howard House, Wagon Lane, Bingley BD16 1WA, UK

First edition 2019

Reprints and permissions service
Contact: permissions@emeraldinsight.com

British Library Cataloguing in Publication Data
A catalog record for this book is available from the British Library

ISBN: 978-1-78769-464-4 (Print)
ISBN: 978-1-78769-463-7 (Online)
ISBN: 978-1-78769-465-1 (Epub)

ISOQAR certified
Management System,
awarded to Emerald
for adherence to
Environmental
standard
ISO 14001:2004.

Certificate Number 1985
ISO 14001

INVESTOR IN PEOPLE

CHAPTER OUTLINE

CONTENTS

Contents

CMF Responsibilities for Implementing Technology-taking 76
Manage the Adoption of Digital-era Technologies 76
Coordinate Adaptation to Digital-era Technologies 78
Challenges of Establishing a CMF 79
Enable Change With a CMF 79

Play 2: Govern Technologies and Change 81
Why Governance Matters 81
Create One Organizational Governance Structure 82
Refer to the Organization's Mission 83
Convene a Governance Committee 84
Establish Real-world, Digital-era Policies and Procedures 85
Develop a Unitary Policies and Procedures Manual 87
Foster Managerial Responsibility for Policies and Procedures 88
Make Managers Policy and Business Owners 89
Empower Business Process Experts 89
Monitor Compliance With Policies and Procedures 90
Train on Policies and Procedures 91
Communicate 92
Analyze Data and Audit Behavior 93
Ready the Organization for Change in the Digital Era 94

Play 3: Engage to Sponsor and Advocate for Change 95
More than Sponsorship Required for Digital-era Change 95
Digital-era Sponsors 97
Change Organizational Culture 98
Sponsoring Actions for the Technology Taker 98
Verbalize the Commitment to Technology-taking 99
Mandate Behavior Changes 100
Suspend Use of Outdated Processes and Require Global Best
Practices 101
Digital-era Advocates 102
Creating Advocacy 102
Advocacy Actions for the Technology Taker 103
Develop a Strong Network in All Functions and Geographies 103
Model the Desired Behaviors 104
Reinforce Adaptation within the Sphere of Influence 105
Engaging Organizations in Technology-taking 106

LIST OF FIGURES AND TABLES

CASE INDEX

PREFACE

WHY READ THIS BOOK

If you are standing on the sidelines wondering how to jump into the digital game, this book is for you. If have seen others pour endless sums of money into failed technology experiments and want to avoid a similar fate, this book is for you. If you are concerned that your organization may be wedded to outdated technologies, this book is for you.

We offer a proven approach for capturing the benefits of new technologies while limiting your business risk. We offer a simple strategy for winning at the technology game, by taking the best of what is available, rather than trying to invent everything yourself. By recognizing that taking on new technologies requires willingness to learn and continually change. We invite you to enjoy the journey.

IDEA IN BRIEF

Digital-era technologies lead organizations to become technology takers, the equivalent of economic "price takers." To be a technology taker is to assent to the behavior transforming benefits of modern technologies. This playbook offers technology takers tactics to manage change, create value, and exploit the digital era's strategic opportunities.

SUMMARY OF THE MAIN ARGUMENT

Users of twenty-first-century digital-era technologies are "technology takers," accepting of and adjusting to whatever the market offers them.

Similar to small firms that lack the market power to set prices and are economic "price takers," managers today are increasingly unable to customize the digital-era technologies their organizations use. Technology takers have little influence over the capabilities of the technologies they adopt; they cannot expect to improve on or customize for themselves the features of Facebook, Google, the iPhone, the blockchain, cloud-based enterprise resource planning systems, or other game-changing technologies.

The inability to modify available information technologies is a shock to leaders and managers alike. Cloud-based technologies arrive with set processes developed by others, and users must learn new ways of working each time the technologies themselves evolve. But refusing to adopt and adapt to digital-era technologies is increasingly not an option. Change in the digital era is constant and behavior-transforming. Leaders must respond to these changes, or they will get left behind by those who do. The constancy of change also means that organizations have to do more than launch typical, one-off change management or transformation projects to succeed.

To adopt efficiently and adapt effectively to behavior-changing technologies, astute leaders should employ change leadership techniques as a strategy for the digital era. This book offers technology takers a playbook to manage change, create value, and exploit the digital era's opportunities. The book draws on research and recent case studies to explain what it means to be a technology taker. Organizations and their managers are offered change leadership plays, which emphasize the iterative nature of change management in the digital era. The book also describes how technology-taking can create value through data stream analytics and be used to respond proactively to the challenges of the digital era.

ACKNOWLEDGMENTS

We sincerely thank our series Editorial Director and Head of Business, Finance and Economics Books, Emerald Group Publishing, Pete Baker. We are grateful to three anonymous peer reviewers of our proposal whose comments helped improve and clarify the purpose and scope of the book. Katy Mathers, Editorial Assistant for Business, Management, Economics, and Finance at Emerald Publishing, is earnestly thanked for her tireless support, as is the cover designer, Mike Hill.

Great thanks are owed to our research assistant, Kathleen Guan, for her copy editing, footnoting, and investigation skills, without which this book could not have been completed. We also appreciate the Latin expertise of classics scholar, Jonas Howard: *scientia est maior aetate*.

Our gratitude is to Claire Messina, Miguel Panadero, Mads Svendsen, Sabine Bannot, Paulo Lyra, and Daphne Moench for support and insights. Joseph Ippolito is also thanked for comments on the initial book proposal, including early foresight into artificial intelligence and related operational strategies. We are grateful to Gerald C. Anderson for sharing his leadership ideas and insights on both strategic and tactical change management tools and techniques.

Further, we are grateful to our families, who have assumed a disproportionate share of domestic duties while we were busy writing. Without the unending emotional support of our spouses, this book would not have been possible.

We are thankful to the support for this book from past and present colleagues. Importantly, the views and opinions expressed in this book are those of the authors' alone and do not necessarily reflect or represent the views of the authors' past or present employers or affiliations. Examples of case materials within this book are examples based on limited and clearly referenced sources in the public domain. Assumptions made within the book are our own and are not reflective of the position of any of the sources cited.

CHAPTER 1

THE TECHNOLOGY TAKERS OF THE DIGITAL ERA

Digital-era information technologies induce organizations to become technology takers, the behavioral equivalent of economic "price takers."[1] In a perfectly competitive market, buyers cannot establish the price they desire for the good they want; these price takers must accept the price offered.[2] Twenty-first-century technologies are beyond the influence of any one organization to customize to meet specific requirements.[3] Organizational leaders become "technology takers," changing their own work behavior to adapt to whatever the modern information technology market has to offer.[4]

Digital-era technologies are ever-changing, frequently updated via cloud computing, and not proprietary or unique to any one organization. This contemporary technology requires two reactions from its users: first, that they adopt the technologies by conducting their work through the technology's processes; and second, that they adapt by modifying their actions use more efficiently the ever-evolving technology.[5] Users exchange their autonomy for efficiency, in that the digital-era technologies used to determine how users must modify their behavior. Most cannot resist using the technological systems that enable modern life; neither can the user refuse the almost daily changes foisted upon her by continuously updating systems.[6] Users must take digital-era technology as it comes and must react in novel ways relevant to the modern age.

THE DIGITAL ERA DEFINED

The fundamental characteristic of current, digital-era technology is that it obligates users' acceptance of its processes and systems. The dominance of cloud-based products gives today's organizational leaders little influence over the functioning of technologies used in the workplace. Failing to change the tech, the users themselves must change.[7] In contrast, twentieth-century technologies were, in most cases, electronic enablers of existing processes, obligating no true, fundamental change of user behavior.

Consider the different behaviors adopted by mobile and smartphone users. The twentieth-century mobile phone enabled conversations via portable, rather than landline, phones. How things were said and how information was accessed did not change. However, smartphones are twenty-first-century digital-era technologies that modify users' behavior because of constant, built-in, system changes. Smartphones access voice, Internet, text, and global positioning data, revolutionizing where, how, and with what information and data streams people work. Smartphone users are technology takers of the options provided by either Apple or Alphabet. Users cannot specify their phones' operating systems, but smartphone use has come to dominate every aspect of modern life, from the time the user awakes to the minute they dim the screen and closes their eyes at night.

Similarly, customized enterprise resource planning (ERP) systems enabled organizations to convert existing business procedures into specified electronic processes. Now, in the digital era, organizations adopting software as a service (SaaS) cannot affect the specifications of the available cloud-based processes.[8] Instead, SaaS defines and constantly redefines shared, globally applicable processes to which users must adapt. Certain SaaS systems, such as Office 365, Dropbox, GoToMeeting, SAP Concur, SalesForce, Workday, and WebEx, have become virtually ubiquitous. From their users, these require constant adoption of their latest process changes and ensuing individual behavior changes to adapt to using these processes. Collaborative technologies also require equal adaptation from all users; as he fumbles to mute his microphone or share his screen, a less-than-fluent user can prevent all others' from having a productive or understandable meeting on WebEx or GoToMeeting.

Like SaaS, data as a service (DaaS) and blockchain as a service (BaaS) too are technology services. Users must use the services' naming conventions and the way the data is broken down; these cannot be modified.

Digital-era technology platforms also require user adherence and conformity. Blockchain is a platform, a distributed digital ledger where transactions are recorded sequentially and publicly. Software firms have developed application software, and programmers have developed open source applications, using blockchain. Companies can also write their own applications using blockchain (including BaaS). These platforms require adherence and are services from a process perspective. Users can apply the blockchain differently but cannot change the way the distributed ledger works.

The digital era also is characterized by the use of and research about artificial intelligence (AI). Python is an AI language that is a technology tool. Using Python does not require adherence to defined business processes, in that an application written in Python is built to specific business process definitions. However, the average user cannot change the AI software embedded in some hardware devices or the AI algorithms in some application software. These must be taken as they come. This AI-enabled hardware and software can be used to control robots. Robotics is a mechanical engineering application of information technology and is the very embodiment of the digital era.

TECHNOLOGY'S PRICE TAKERS

The relationship of digital-era technologies and technology takers is similar to the reactions of small firms in a globalizing marketplace. The transactions of small companies and individual consumers are unable to affect the market price of a good. The price is set by the greater forces of supply and demand. Businesses must accept the prevailing prices in the market for the sale of their products, and they must distinguish their products in some other way than price. Small firms are price takers and obtain profitability through decreasing production costs, increasing the volume of sales, or through some other internal effort.

The economic model of perfect competition, which leads to price taking, makes several assumptions that can be analogized to the behavioral model of technology-taking.[9] In a perfectly competitive market, goods are identical and cannot be distinguished from one another. The market has a large number of buyers and sellers; so many, in fact, that none can affect the market price. Although the perfect market already has many competing firms, more businesses may enter or exit the market at any time. Finally, the perfect competition model assumes that each player in the market has complete information about the market's prices and operations and that information costs little to obtain.

Technology-taking in the digital era is similar to the ideal market underlying the microeconomics of price taking. In the perfect competition market model, there are many buyers and sellers, and the products offered tend to be quite homogenous. Similarly, smartphones, whoever their maker and regardless of whether based on systems by Apple or Alphabet, have flooded the market and tend to be very similar in product scope. Digital-era applications are globally applicable processes available for sale to organizations (SaaS) or at zero cost for the general public (Facebook, Instagram, Snapchat, Google, etc.). Low general or relative cost makes consuming digital-era technologies exceedingly easy to do. For example, SaaS-based ERPs are comparatively cheaper than their alternative, customizable ERP options, because the buyer of customizable systems has ever increasing update and maintenance costs.

In a perfectly competitive market, there is low cost of entry and exit for price takers. Absent regulatory restriction, anyone can set up shop selling vegetables or cooked food or widgets; and, if the business is not profitable, it can be closed. Analogously, twenty-first-century technologies that drive technology-taking are rarely proprietary or restricted to one user or organization at the point of adoption. And some digital-era technology is offered free at the point of consumption, as is the case with Amazon, Google, and Facebook applications. Virtually every modern consumer technology is based on the Internet, itself with very low costs of entry.

Small firms cannot opt out of the marketplace, for the market is the ecology in which the firms operate and find their customers. There is no alternative. Users of smartphones and SaaS too cannot decide to use only some of their technology's operating systems. To accept a part is to accept the whole ecosystem of an iPhone or a cloud-based ERP; both require the full adoption of the product offered. Yes, a user could never open, say, the mapping application on her smartphone; but, it is always there in the background, its global position system enabled and its data stream uploading to the cloud. The minute the user accesses Facebook or Google via her smartphone, these systems obtain the user's geolocation data from her smartphone to create a bespoke Facebook or Google experience.

Arguably, the information about the price and functioning of digital-era technology is offered at low cost to all interested users.[10] Applications on smartphones are free or cost pennies and have made widely available smartphones' almost endless capabilities. Information is amply available about modern tech, often through the very cloud-based systems on which these technologies rely.

RELUCTANT TECHNOLOGY TAKERS

The technology taker concept works in different types of markets with different levels of competition. For some digital-era technologies, the taker is forced into that position by monopolies or oligopolies of the technology. Prices (or our technology-taking analogy) are set either because there is much choice or because there is too little choice to meet demand. Where there is lots of demand and few options, the price (or preferred and possible technology offering) is set artificially high: a singular technology option is restricted to those able to access and afford it. The marginal effect of one more user demanding a technology is nil, but for a different reason than in the perfectly competitive market. In monopolies, the tech maker would determine the functioning of the market, because demand would not matter at all.[11]

Google is a near-monopolist of Internet search, but one that has decided on a price point of zero. So too is Facebook a monopolist of social networking with no price. The user of these technologies takes them as they come and has little influence on their offerings. Users may not have contemplated the true costs of using the "free" services of Google or Facebook or Amazon marketplace. The cost extracted by digital-era technologies, where, moreover, the perfect competition model of tech meets reality, is that these technologies require users to change their behaviors.

TECHNOLOGY-TAKING AND THE BEHAVIOR CHANGE DELTA

Technology-taking requires constant behavior change of users of digital-era technologies. Behavior change implies an explicit need to manage that change. As defined for the digital era, change management is the acquired organizational skill set of dealing with the entrenched inability of managers and their organizations to recognize the need to (1) adopt, rather than resist, the technologies that are now fundamentally changing entire industries, and (2) adapt behaviors, rather than customize the technologies.

Those in denial of the digital era's demands put people, processes, and technology on equal footing when implementing change. In classic, project-level change efforts, people or processes would drive change. Technology was at the bottom of the Behavior Change Delta, not at its apex, because there was no harm to a process or people-first approach. Technology building blocks were seldom sufficient for business transformation, certainly did not cause business transformation, and their lack seldom held a business back.

In the digital era, technology takers must recognize that technology is now at the apex of the Behavior Change Delta. Because of constant, technology-driven changes, the processes workers must use are increasingly becoming subsumed by technology. Now, the intersection among technology, people, and processes is the focus when managing behavior change (**Figure 1**).

As faculty and advisors to managers across industries, we have used string to lay out the Behavior Change Delta areas of process, people, and technology on the ground. We then ask students to step into a particular triangle to indicate answers to two questions.[12] First, we asked them to

---

Figure 1: The Behavior Change Delta.

step into the triangle where their past and present organizational change projects had taken place. Most stepped into the process triangle; a few lurched, almost apologetically, into the technology triangle.

Then, students were asked to move into the triangle they considered most important for an organization undertaking a change project to support the organization's mission. To answer this question is challenging from a change management perspective. The answer also indicates an organization's ability to deliver on its mission while remaining successful in the digital era. Usually with a sense of relief, most students stepped from process to people. A few stayed in the process delta, but rarely did anyone stay or move into the technology triangle.

Clearly, the idea of technology-taking has not fully caught on. Prior to the digital era, strategy work, as well as change management, was focused on people and process. Many change management projects tried to ensure the hiring of people with skill sets that matched those demanded by customers. Or change was planned around either strategic or tactical actions driven by a marketing or a production method or a new set of strategic goals. Or internal processes were the starting point for customizing an ERP. In these examples, people and process changes would be enabled by technology in support of the expected or resulting behavior changes, which in turn were calibrated to support the organization's mission or strategic intent.

Now, technology has ascended to the top of the Behavior Change Delta. Twenty-first-century technologies are not merely replicating or enabling existing processes or ways of doing things; they are forcing changes in standard/grandfathered practices. It would be hard to find a chief information officer from the 1990s who would have predicted the death of formal, validated business requirements and the rise of a technology-first adoption process.[13] Organizations and their leaders must adopt digital-era technologies that are interrupting entire industries – from retail (Amazon) to political campaigning (Facebook).

Further, people and organizations must adapt their workplace behaviors to these technologies, and adaptation to digital-era technologies is constant and never-ending. Technology takers are in an iterative game because the technologies they use are constantly being updated and revised. Change management, including leaders' sponsorship of change, too must be continuous and geared to the long term.

Digital-era technologies are beyond the influence of any one organization. Instead of accepting and managing the inevitable changes of the digital era, many organizations are still vigorously trying to fight them. These organizations insist on redefining, re-engineering, or rejiggering internal processes. But no one user has control over the algorithm of Google Search and would be somewhat silly to try unilaterally to improve it. It is the dominant search technology for its ease of use and comprehensiveness. Yet some libraries still argue that their own, dedicated search engines provide better academic research facility than Google.[14] The proof is in the virtual pudding: Google controls 72% of the search engine market share, with other search engines not even coming close to this level of use.[15]

The futility of process re-engineering in the digital era results in the loss of tangible space for management. Previously, managers were experts in the proprietary processes they led. Now, organizational processes have been supplanted by global norms dictated by the technology used. Managers may be highly resistant to the behavior changes demanded of them because these require a wholesale reinvention of workplace roles and responsibilities.

This resistance must be managed. Technology takers focus on strategy and commit to using technology to avoid being run out of the market by other companies using the same technology. They then support their organizations to match their processes with those required by the technologies used. Technology takers realize that using digital-era technologies will require behavior changes that will put the organization in a better position to create value.

CHANGING BEHAVIOR

To be a technology taker is to assent to behavior change. Facebook offers a free service for billions of individual users. In exchange for no-cost access to social media, the user consents to Facebook's sale of the individual's data stream to advertisers. To opt out of Facebook is to opt out of access to the social marketplace. Instead of losing entry to the online world, Facebook users have altered their behavior and expectations to accommodate Facebook's requirements.

Blockchain is another example. This technology harmonizes all users' behavior because blockchain offers no ability for users to change the virtually linked servers that are the basis of the technology. If they seek to use the blockchain cost reduction, security, and efficiency features, users must fully adopt all systems and requirements of the blockchain. This condition for using blockchain is acceptable – attractive, even – to industries as diverse as banking, shipping, and nonprofits like the World Food Programme (WFP), which has adopted a blockchain distributed-ledger model for food programs for refugees.[16]

Even McDonald's, famed for unique, proprietary processes, has had to change its organizational behavior for the digital era. McDonald's has started eliminating its over-the-counter and drive-up window ordering systems, internal technologies developed over decades. Instead, McDonald's restaurants increasingly direct customers to order from self-service, smartphone look-alike, interactive screens.[17] Customers are familiar with these screens due to the global dominance of smartphones' processes. McDonald's customers too can use self-service ordering technology to craft their hamburgers according to their wishes – with extra ketchup and hold the pickles, please. This, in turn, has led to a true revolution for McDonald's, a company founded by Ray Kroc to be the assembly line of standardized hamburgers, all alike with exactly two pickles per burger.[18] Modern customers now demand choice, and McDonald's has used technology-taking to deliver choice.

VALUE CREATION IN THE DIGITAL ERA

As with all else that has changed in the new era, organizations' value also is generated in novel ways enabled by new technologies. The purpose of managing change is to ensure that whatever changes foisted upon an organization enhance – and do not detract from – the organization's mission and help align mission delivery across the organization. Change management should seek to increase the value created through technology use in support of the organization's vision and mission: what the organization is trying to accomplish, its mode of operation, and where it operates.

New technologies offer value creation opportunities through information gained from data streams. Cargill, the grain distributor, recently invested in a "data refinery" to clean and process data about predictive crop yields into usable, actionable information.[19] Already the largest privately held US corporation, Cargill can add value to its bottom line by selling its data analysis to farmers for weekly production forecasts and to other companies involved in shipping, forestry, and energy.[20] Cargill's potential profits from data analytics could be astounding if others' success in trading in data is any indication. Facebook, a corporation that has sold to advertisers its analysis of its users' data, posted a US$4.26 billion profit in 2017, a 61% increase in profits over the prior year.[21]

For example, a rider is a technology taker of the digital-era process of taking an Uber or Lyft. These transportation companies have defined a globally applicable process for hailing a private taxicab. Adapting to that process allows the rider to use Uber or Lyft data to improve her own ride experience and analyze cost; in turn, those data are shared with millions of others to improve their own transportation planning. The information in the technology takers' data stream is what matters, not the process by which it is generated.

When organizations use globally created, rather than locally or internally defined, processes, technology-taking drives the implementation of best practices and the change of workplace behavior for increased productivity. Organizations should manage the changes wrought by technology-taking. Leaders' change management goal should be value creation through analysis of mission or effectiveness related information, actively incorporating data streams into their strategic planning frameworks. If nothing else, they should ensure that they use data streams for strategy formulation metrics. To survive and thrive in the digital era, organizations must employ continuous change management to drive work practice efficiency and data-driven effectiveness in the achievement of organizational goals.

CHANGE MANAGEMENT PLAYBOOK FOR TECHNOLOGY TAKERS

Change management ensures that the new processes resulting from a project are actually adopted by the people who are affected. Leaders need to communicate what is constant – values and mission – and what is changeable – processes and behavior. Organizations must build a culture that permits innovation and transformation. Technology-taking leaders and organizations seek opportunities in a constantly changing environment by recognizing the value-creating benefits of the digital era. As revealed in our playbook, change management plays include envisioning, governing, engaging, equip, and measuring to match the ongoing, rather than one-off, transformation challenges presented by the digital era. Each of these plays must be responsive to the new challenges posed to sometimes unwitting or reluctant technology takers.

Today's leaders lack the ability to customize the technology used by their organizations, while digital-era technologies drive changes in management techniques, processes, work behaviors, and organizational language. These changes are challenging because proprietary practices and processes developed across decades are, by definition, no longer best practices. Many middle managers have built their careers on understanding the quirky, unique processes of their organization. Any change threatens their legacy of knowledge.

A change management playbook offered for the digital era is therefore helpful to those who want to lead their organizations' adaptation to digital-era technologies. Cloud-based, modern, technological change is ceaseless, making constant the need for change management for any organization that wants to succeed in the digital era. Based on our research and experience with digital-era technologies, we recommend technology takers rely on the following five plays to lead change in their organizations.

Play 1: Envision — Create a Change Management Function

So that it can be managed, what is changing must be identified and organized. With a clear leadership vision of technology-taking established, a change management function is required to perform planning duties at the strategic level to ensure follow-through and measurement of technology-taking successes in the wider organization. In the digital era, change management is not a purely voluntary endeavor or a "nice-to-have" function for creating a project-by-project urgency. The constancy of change makes change management vital. Change management guides the organization to link its mission and goals with its adoption of technologies and technologically induced behavior changes.

Creating a change management function (CMF) is an indispensable play to spearhead the effective adoption of digital-era technologies and to move leaders' focus from process to data analytics. This gives credible support to leaders in setting out a technology taker strategy for the organization. The organization's change management function can review available technologies, for example, novel applications of blockchain and tools for data stream analysis for better decision-making. The CMF can help the organization focus dispassionately on the data generated, not the process itself, to make decisions and craft solutions to specific business problems.

Play 2: Govern – Establish Governance of Technology Adoption and Adaptation

The change management function will also support improved managerial response to constant change by setting up governance structures to coordinate and authorize technology adoption and adaptation. Governance is a coherent framework of properly documented policies and procedures. These reflect decisions to guide the organization on how to adopt and adapt to externally defined, digital-era technologies.

Investing in overall organizational governance is crucial, as there is usually fierce resistance to changes to managerial power and decision-making. Integrating technology governance within an organization's governance structures allows coordination of change beyond functional areas. Silos of perceived expertise often are fiefdoms of mid-level managers who are most likely to resist technology-taking. Without comprehensive governance covering technology use and data analysis, these managers will continue to try to customize systems to their own needs instead of adapting to digital-era processes. Customized systems may optimize results for one or two functional areas but will be suboptimal for the organization as a whole.[22]

As part of modern, revitalized governance for a digital-era organization, new roles and responsibilities should be given to people responsible for implementing, explaining, or leading change. The organization should create a cross-functional cadre of business process experts (BPEs) who are accountable for horizon-scanning for better solutions and describing how adopted processes generate business data and create value.

Play 3: Engage — Sponsor and Advocate for the Constancy of Change

Classic literature on organizational change management posits that managerial-induced transformations lead to more efficient and effective organizational cultures. As the lynchpin of managing change, an organization's leaders are to create a Kotterish "sense of urgency" for these changes.[23] Other critical success factors like executive buy-in are imperative.[24]

Sponsoring change is thus an essential step in leading digital-era change management. "Sponsoring" means welcoming change with open arms, explicitly communicating about forthcoming changes, and endorsing the idea of constant change. An effective change leader will sponsor the constancy of change, which can make way for the adaptation of behaviors instead of systems.

Nonetheless, the digital era requires more than sponsorship alone for the successful adoption of and adaptation to its technologies. Advocates too must lead their colleagues to change by serving as role models. Advocacy combines agency and championship actions used in other contexts to convince allies of a cause; and, it can be applied to technology-taking. Together, advocates and sponsors can engage their colleagues in productive communication and action for change.

Play 4: Equip — Train Technology-taking

Some purveyors of digital-era technology argue that their technologies are so intuitive that in-person training is not required for organization-wide adoption of their technology applications.[25] Using Google, the leading Internet search provider, involves no training. However, most digital-era technologies represent a significant change in work practices; and, training people on the use of these technologies will help instill desired workforce behavioral changes. Many people lack the skills needed for the digital era, including the ability to draw conclusions from data. Through training, organizations equip their people and teach them the new skills required for the technologies adopted. Astute leaders also will use training opportunities to communicate, acknowledge fears and frustrations, and invite people to create their own technology-taking future. Training too will describe and model desired behaviors to reap the gains available from technology-taking.

An erroneous assumption is that workers will self-learn technology. Training must be a proactive exercise, placing everyone in an organization, from workers to middle managers to executive leadership, in situations where they can communicate and benchmark themselves to improved digital-era performance. Training helps people stay proficient using ever-changing technologies, and it establishes benchmarks against which managers can measure the depth of the workforce's acceptance of change.

Play 5: Measure — Evaluate Managers' Embrace of Technology Change

Modern leadership frameworks supposedly break down silos, helping people embrace new ways of working and making transparent the organization's commitment to change.[26] Digital-era leaders must embrace the complexity of behavioral change propelled by technology-taking and build a culture accepting of constant change.

If an organization has committed to technology-taking, its managers must be measured by their technology adoption and adaptation. All senior management, including legal, finance, and executive officers, should be tested on technology use and awareness. If their technological abilities are not reviewed, some managers mistakenly will believe they can opt out of being technology takers. They may continue to put people, processes, and technology on an equal footing, a mistake given that constantly changing technologies have subsumed the processes workers must use.

Leadership involves managing the interaction between technology and people with clarity on the strategic intent. Data as a service (DaaS) can be used to establish this and other benchmarks of effective leaders and for measuring an organization's leaders against these benchmarks.[27] Holding executive management accountable may imply replacing leadership that has been unable to evolve with the digital times.

LEADING CHANGE IN THE DIGITAL ERA

Technology-taking is both an emerging trend and a tide that forever sweeps along all organizations, managers, and workers. Although adopting the technological practices of others and adapting workplace behavior to accommodate these technologies may be unwelcome or controversial, the technology-taking trend is futile to resist. The digital era is our modern reality and is remaking all industries and entire societies.

Most organizations have already confronted the digital era, but with varying success. The prevailing winds within many organizations continue to push them toward customized technological solutions, wherein an entire computing system is created for a single organization or where a system is so configured for one company that it loses its commonly available business practices. The failure rate of these customized systems is legendary.[28] Yet, even with specialized systems' reputation for cost overruns and catastrophe, organizations continue to try to reinvent technology for themselves. Instead, organizations should accept that they are technology takers and determine how to use technology to increase organizational value.

For those recognizing that we are in the digital era, we offer you technology-taking as a strategy with a practical change management playbook. We offer you ideas on how to create value through data analysis and behavior change, which can then be implemented through our playbook to digital-era change leadership. Being a change leader in the digital era allows you and your organization to flourish through a virtuous cycle of change.

NOTES

1. We define digital-era technologies as ever-changing, frequently updated via the cloud, and not proprietary or unique to any one organization that uses them. The "cloud" refers to remote data centers that host computer memory, processing power, and applications ("apps"). Digital-era technologies include those that are "smart," in that they know who you are, where you are, what you are looking for, and how to pay for it. See Regalado, A. (December 30, 2013). *Who coined 'cloud computing'?* Retrieved from https://www.technologyreview.com/s/425970/who-coined-cloud-computing/and Pharoah, M. (April 9, 2018). *Transforming change management with artificial intelligence (AI).* Retrieved from https://www.andchange.com/transforming-change-management-artificial-intelligence-ai/

2. University of Minnesota, *Perfect competition: A model.* Retrieved from https://open.lib.umn.edu/principleseconomics/chapter/9-1-perfect-competition-a-model/

3. Avila, O., & Garcés, K. (2016). Change management support to preserve business–Information technology alignment. *Journal of Computer Information Systems, 57*(3), 218–228. doi:10.1080/08874417.2016.1184006

4. For application of the price taker concept to political science, see Heng, Y., & Aljunied, S. M. A. (2015). Can small states be more than price takers in global governance? *Global Governance, 21*(3), 435.

5. We use Hall and Kahn's definition of technology adoption: "The choice to acquire and use a new invention or innovation." Hall, B. H., & Khan, B. (November, 2002). *Adoption of new technology.* Retrieved from https://eml.berkeley.edu/~bhhall/papers/HallKhan03%20diffusion.pdf

6. Nicas, J. (April 01, 2018). *They tried to Boycott Facebook, Apple and Google. They failed.* Retrieved from https://www.nytimes.com/2018/04/01/business/boycott-facebook-apple-google-failed.html

7. Desmet, D., Löffler, M., & Weinberg, A. (September, 2016). *Modernizing IT for a digital-era.* Retrieved from https://www.mckinsey.com/business-functions/digital-mckinsey/our-insights/modernizing-it-for-a-digital-era?cid=eml-web ("The sheer volume of technologies, processes, and decisions required to build and maintain digital applications and operations means companies can't afford to work in the same old ways").

8. "If a company's IT department is not capable of moving fast enough, any other department now can go online, find a software-as-a-service application and provision that application themselves relatively easily. Then the C-level technical executive and his people are left to figure out how to manage all of these new demands and devices, and monitor and maintain the user-provided infrastructure": Beckley, A. M. (August 07, 2015). *How the cloud is changing the role of technology leaders.* Retrieved from https://www.wired.com/insights/2013/09/how-the-cloud-is-changing-the-role-of-technology-leaders/

9. University of Minnesota, *Perfect competition: A model.* Retrieved from https://open.lib.umn.edu/principleseconomics/chapter/9-1-perfect-competition-a-model/

10. Graham-Harrison, E., & Cadwalladr, C. (March 17, 2018). *Revealed: 50 million Facebook profiles harvested for Cambridge Analytica in major data breach*. Retrieved from https://www.theguardian.com/news/2018/mar/17/cambridge-analytica-facebook-influence-us-election

11. University of Minnesota, *The nature of monopoly*. Retrieved from https://open.lib.umn.edu/principleseconomics/chapter/10-1-the-nature-of-monopoly/

12. Thanks to Sabine Bhanot and Daphne Moench for showing us this technique.

13. Andriole, S. (April 13, 2018). *Implement first, ask questions later (or not at all)*. Retrieved from https://sloanreview-mit-edu.cdn.ampproject.org/c/s/sloanreview.mit.edu/article/implement-first-ask-questions-later-or-not-at-all/amp

14. Lessick, S., & Kraft, M. (2017). Facing reality: The growth of virtual reality and health sciences libraries. *Journal of the Medical Library Association, 105*(4), 407–417. doi:10.5195/jmla.2017.329

15. Comparing use of Google with 10 other search engines in the 12 months from June 2017 to June 2018: Net Market Share. (n.d.). *Search engine market share*. Retrieved from https://www.netmarketshare.com/search-engine-market-share.aspx

16. World Food Program. (n.d.). *Building blocks*. Retrieved from http://innovation.wfp.org/project/building-blocks

17. Gavett, G. (August 04, 2016). *How self-service kiosks are changing customer behavior*. Retrieved from https://hbr.org/2015/03/how-self-service-kiosks-are-changing-customer-behavior

18. Kroc, R., & Anderson, R. (1987) *Grinding it out: The making of McDonald's*.

19. Cosgrove, E. (August 24, 2017). *Cargill invests in predictive Ag 'Data Refinery' Descartes Labs' $30m series B*. Retrieved from https://agfundernews.com/descartes-raise.html

20. Ibid.

21. Rushe, D. (January 31, 2018). *Facebook posts $4.3bn profit as Zuckerberg laments 'hard year'*. Retrieved from https://www.theguardian.com/technology/2018/jan/31/facebook-profit-mark-zuckerberg

22. But see Beckley, A. M. (August 07, 2015). *How the cloud is changing the role of technology leaders.* Retrieved from https://www.wired.com/insights/2013/09/how-the-cloud-is-changing-the-role-of-technology-leaders/ ("The cloud era is driving businesses to recognize the importance of software-as-a-service that is configurable to the organization's own particular needs and processes, rather than having to adapt their processes to some standardized system").

23. Kotter, J. P. (1996) *Leading change.* Boston, MA: Harvard Business School Press.

24. Luo, J. S., Hilty, D. M., Worley, L. L., & Yager, J. (2006). Considerations in change management related to technology. *Academic Psychiatry, 30*(6), 465–469. doi:10.1176/appi.ap.30.6.465

25. Touting Workday's "on demand and on the go" training modules – on only social media and video. (2017). Workday for financial services. *Workday.* Retrieved from https://www.workday.com/content/dam/web/en-us/documents/datasheets/datasheet-workday-for-financial-services-us.pdf

26. See Digital-era change runs on people power. (August 09, 2017). Retrieved from https://www.bcg.com/en-cl/publications/2017/change-management-organization-digital-era-change-runs-people-power.aspx

27. Press Release: Workday delivers its first data-as-a-service offering with workday benchmarking. (October 10, 2017). Retrieved from https://www.workday.com/en-us/company/newsroom/press-releases/press-release-details.html?id=2190890

28. Fruhlinger, J., & Wailgum, T. (July 10, 2017). *15 famous ERP disasters, dustups and disappointments.* Retrieved from https://www.cio.com/article/2429865/enterprise-resource-planning/enterprise-resource-planning-10-famous-erp-disasters-dustups-and-disappointments.html

CHAPTER 2

TECHNOLOGY TAKING AS A STRATEGY

Few business leaders have the inclination, capability, or resources to write apps for their smartphones. Consequently, they have a take-it-or-leave-it strategy with respect to cell phone apps for personal use. They do not download apps and then modify them. They do not create apps from scratch. Instead, they choose which apps to use and unhesitatingly shape their behavior to take advantage of the features. They explore, they learn, they have fun! Many of these same leaders have not recognized that a similar strategy of technology-taking is now available for business applications.

Technology-taking is a new concept as the world moves into the cloud of digital-era innovation. New strategic possibilities are offered by the shift in technology from merely enabling to behavior changing. We propose the novel perspective that technology-taking is a strategy that permits an organization to succeed in the digital era.[1] Technology-taking organizations will commit to using digital-era technology, even as it increases competitive interconnectivity with external stakeholders; match their processes with those required by the technologies used; and require changes in behavior so that management decision are based on information from data streams. In exchange for these efforts, the digital era will reward organizations with new ways to create value, perhaps through gains in efficiency or perhaps via the discovery of unique contributions to the modern market.

Very few leaders have internalized the profoundness of the shift from the enabling era to the digital era, for this shift occurred so quickly. By "enabling," we refer to computerized technologies, the use of the Internet, and the storing of organizational information on an organization's own servers. Enabling also implies that leaders can always choose the technology that is right for them, that technology responds to organizational needs, and the organization needs, not adopt and adapt to what the market has to offer.

As soon as many organizations adapted to a world enabled by computers, a giant leap occurred. Now, the digital era is characterized by behavior-changing, cloud-based computing, continuously updating systems, big data, artificial intelligence, and a reduction in the usefulness of proprietary systems and information. Technology offers up a brave new world where organizations do not waste resources by trying to replicate processes that are no longer contemporary or that are based on failing business models.

From a strategic perspective, those changes compare in scope to the "almost miraculous improvement in the tools of production" of the industrial revolution.[2] We now can work anytime, from anywhere, with access to almost any data to compare, manage, and generate value for ourselves and our organizations. But then as now, radical change came with significant cost. The industrial revolution was "accompanied by a catastrophic dislocation of the lives of the common people."[3] This dislocation was similar to that experienced by the cavalries of modern armies, for which the mechanization of war also came quickly and unexpectedly. Armies' use of the calvary continued into the era of trucks, tanks, and airplanes, often with disastrous results. Horse-riding soldiers had their last stand in twentieth-century warfare.[4]

DIGITAL-ERA DISLOCATIONS AND DISINTERMEDIATION

The digital era too has caused both short- and longer-term business model dislocations.[5] The examples are plentiful: Kodak's slow-motion implosion because of a failure to scale its own digital technologies,[6] Blockbuster's annihilation by Netflix,[7] Disney's recreation of Netflix,[8] the consolidation and reinvention of white-shoe law firms,[9] and lawyers' predicted replacement with artificial intelligence.[10]

In recent years, the concept of disintermediation has also gained currency in more technical discussions about who is getting rewarded from the digital era's business model dislocations. Disintermediation is defined as the elimination of the intermediary in a transaction between two parties.[11] Technological disintermediation occurs when an interaction is more directly made through the use of behavior-changing technology. The haberdashers, the IT departments, and the home video purveyors have been disintermediated. The modern world has no use for them, for we no longer wear hats, require on-site tech support, or use video cameras. We instead wear Google glasses, send online help tickets, and use our iPhones.

As technology replaces the middleman, it also changes the behavior of the still interacting parties. Groceries are still bought, even if they are purchased using technology that disintermediates the need for a brick and mortar store. The Internet changed the intermediary role which the advertising agency traditionally occupied between advertisers and the media.[12]

As noted by Mintzberg, when organizations "understand the difference between planning and strategic thinking, they can get back to what the strategy-making process should be: capturing what the manager learns from all sources [...] and then synthesizing that learning into a vision of the direction that the business should pursue."[13] Transposed to ensure technology-taking, the direction organizations should pursue is that toward the digital-era future and beyond. That requires a readiness to change.

Driverless trucks are not yet on the road in North America but are soon to traverse the highways of Canada and the United States.[14] These may disintermediate truck drivers from their jobs, as humans are replaced with artificial intelligence. Or they may create more attractive jobs for human truckers by relieving the stress and monotony inherent in long-haul driving. Trucking jobs may be transformed into work at remote control centers or control functions similar to those of airline pilots or tugboat operators. Trucking companies' strategy now needs to be ready to anticipate the future, engage its stakeholders to mitigate dislocation for truckers, act upon the information and interconnectivity created from driverless transport, and measure itself against goals apt for the digital era.

MAKER, TAKER, TINKER, AND TAILOR: THE ADOPTION–ADAPTATION STRATEGY MATRIX

The digital era requires leadership and a renewed strategic approach to adopting continuously renovating technologies and adapting behavior to changing technology. Organizations' leaders must plan to stay ahead of the technology curve, avoid reliance on faltering business processes or plans, or resist being disintermediated from the market and going out of business altogether. *The Adoption–Adaptation Strategy Matrix* shows that players in the digital era can choose among four strategies of varying amounts of adoption of and adaptation to technologies. One can be a Maker, Taker, Tinker, or Tailor. The quadrants of the Strategy Matrix are outlined as follows (Figure 2).

The first quarter is the *Technology Tailor*. Those who choose this strategy approach adaptation with an outdated mindset – and with little desire for behavior change. Tailors adopt technologies and then seek to customize them according to a detailed set of requirements. Those specifications are known only to the organization itself and are used exclusively as internal business support tools. As the digital-era marketplace changes, Tailors must continually recustomize their processes in response. The organization becomes ever disconnected from the latest technologies available. Adopting the latest technologies becomes challenging, and upgrading existing systems cumbersome.

For the Technology Tailor, internally defined business processes are of paramount importance and must be protected from change. Tailoring is often held out to be a low-risk strategy, with old processes preferred over adapting to the requirements of new technologies. Organizations that have

Figure 2: The Adoption–Adaptation Strategy Matrix: Maker, Taker, Tinker, and Tailor.

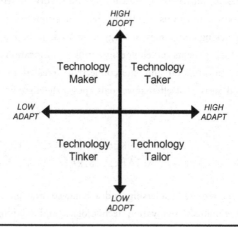

chosen the Technology Tailor Strategy claim that they should specify what technology the organization requires for its success.

Yet technology customization has hidden costs. Similar to a poor-fitting, overtailored suit, tailored technologies do not fit in well with the digital era. These technologies are proprietary to the organization and do not update automatically. Regardless of the initial cost and splendor of the technology tailoring job, this technology will soon fall out of fashion. As the body of business changes, organizations must pour money into their customized technologies to conform them to whatever needs are current.

The *Technology Tinker* wishes neither to adopt digital-era technologies nor to adapt themselves to these technologies. Tinkering is both the conscientious choice of avoiding contemporary technology as a tool of enlightenment and a business-as-usual approach. The Amish have for centuries practiced conscientious resistance to both technology adaptation and adoption. A successful agrarian community, at least in the United States, the Amish still rides in horse-driven carts and eschew electricity and other modern conveniences. In-the-cloud technologies certainly are not the Amish's concern.

The Technology Tinker Strategy is not uncommon for other established businesses, like the taxis in Washington, DC. These businesses may use some technologies, but those technologies are of the past. Neither tech nor behavior has been updated for the digital era. In DC taxis, the rider sometimes has to beg to pay with a credit card, and rarely will the driver permit non-cash payment. Most Technology Tinkers will be disintermediated by the digital era. In fact, the effect of digital-era ride-sharing services has been catastrophic on the DC taxi business.[15]

Organizations can also strive to be *Technology Makers* of technologies for others' adoption and adaptation. Makers like Uber, Facebook, Apple, Google, etc. invented the technology they have adopted. But makers do not need to adapt their behaviors to outside technological forces; they own these technologies and can deploy them at will.

Some Technology Tailors have a mistaken belief that they are Technology Makers. But investment in excessive customization and re-engineering does not a Maker make. Customization is not equivalent to making available to the market configurable and widely adoptable and adaptable technologies.

This book focuses on the *Technology Taker Strategy*. Technology Takers adopt globally applicable, externally defined processes and change their behaviors to adapt to using technologies. Given the difficulties of adopting and adapting, Takers use change management techniques to address the digital era's challenges and opportunities.

The agricultural commodities distributor Cargill recently was forced to change its business strategy to one of technology-taking, rather than being left in the digital dust like Kodak or Blockbuster. Instead of only buying and reselling corn, Cargill now uses cloud-based technologies to provide "big data" to farmers. The *Corn and Soybean Digest* reports that Cargill realized, "You don't have to adopt new technologies, but you have to compete against those who do."[16] Anticipating, using, and responding to new technology separate winners from also-rans.[17]

The Technology Taker Strategy is a plan of action that recognizes that the organization has decided to use constantly changing technologies defined by others. The use of cloud-based process means there is no end to technology-taking and no time where an organization has fully implemented all technology-driven changes. To play in the digital era, an organization builds into its strategic planning the idea of adopting and adapting to constantly renovating, externally controlled technology. A technology taker plans with data as the base of decision making and positions her organization for increased interconnectivity as technologies evolve. Technology-taking as a strategy is used to ensure that digital-era technology furthers the vision and contributes to the successful delivery of the organization's mission.

TECHNOLOGY TAKER STRATEGY GUIDING PRINCIPLES

A set of guiding principles can complement and help remedy deficiencies in digital-era strategic planning; ensuring strategic implementation remains congruent with technology taking. Organizational leaders should lead the envisioning part of how technology-taking will work for their organizations, how it can help further the organization's mission. Leaders must also sponsor those strategic objectives that support digital-era readiness and engagement with managers and people in their organization. When choosing among options of how to proceed, the digital-era option should be the preferred course of action. Technology should be in balance with an organization's people and procedures, no matter the importance of the person or her desire to control unique processes. Management decisions should be based on the data generated by these technologies. Strategies based on the following guiding principles will vary across organizations and industries, but a commitment to these principles can ensure that any strategy developed is a technology-taking one.

Principle One: Consider Technology-taking as the First Option

Technology takers should adhere to a guiding principle of preferring that digital-era technologies are allowed and encouraged to influence both strategic planning and strategy implementation. The digital-era technology should be the preferred option, for it is the more efficient and effective alternative to processes that have been left in the technological past.

To illustrate, any millennial is instinctively familiar with sharing to Instagram a photograph snapped on a smartphone. But for people who grew up with Kodak film that was developed and printed, the behavior change component of sharing pictures may have escaped them. The digital camera made electronic the process of taking photos: to show the pictures to others, the photographer needed to take the SIM card out of the digital camera, put it into the adapter, and then connect the adapter to the TV for the viewing of anyone close enough to the TV screen.

Instantaneous sharing of photos with all the world may not be among the Kodak generation's perceived available options. Yet the world marches forward. Technology takers must presume that the wet signed memo should be replaced with electronic signatures, the executive order with an approved virtual process, and the email with instant communication within a social media platform or ERP.

Principle Two: Position Technology Above People and Process

Being a technology taker is to put technology above processes and people, as the primary source of change to benefit the organization. The reason for taking action to promote technology in the Behavior Change Delta is the risk of digital-era disintermediation, where technology will replace the work or worker. An organization that needs a process enabled avails itself of the technology developed by a maker. Sometimes, the organization moves itself from being a technology maker to a taker – voluntary disintermediation in favor of technology-taking. The increased adoption of SaaS shows this trend. For organizations, SaaS is a less risky option than traditional ERPs because SaaS costs less to implement and is more likely to be implemented on time.

Other times, an organization is disintermediated by others who have become technology takers where it has not. Those are the cases where the organization, for example, Blockbuster or taxi drivers, is forced out of its business through digital-era technologies. Ride-sharing and content-streaming companies have disrupted the transportation and home entertainment industries. The technology was the impetus for change, not process or people.

Many organizations are unable to differentiate between technology-taking and making. Most have tried to be technology tailors or makers, putting their own business processes and people requirements first and then attempting to make technology enable those requirements. In rare cases, an organization will decide to be a technology maker for competitive reasons or because the available technology does not meet their needs. But in the digital era, technology making is likely the exception, not the norm. Sharing processes is more cost effective and provides access to better data stream options, which in turn may lead to better long-term business value.

Principle Three: Base Management on Data

Technology-taking strategic planners must lead their organizations to analyze data and obtain the maximum potential of digital-era technologies. Most organizations do not focus beyond profit and loss statements, sales pipelines, multiyear indicators, and the like, which will not allow them to align transactional data into analyzable dashboards.

Digital-era technologies generate data that organizations can use to inform decisions and to increase value. But for this data to have worth and power, organizations must invest in data analysis and commit to using that analysis to affect strategy. To use data for strategic planning purposes is to analyze data to determine whether, when, and how an organization has reached its goals. Digital-era technologies provide organization's data from business processes, of a continuous nature, and in real-time. This information can be used to test hypotheses and measure progress toward strategic objectives.

For example, Scania, the truck maker, has expanded its business strategy to collect and analyze data from each of its trucks to increase business efficiency. "Transport is becoming a data business," and Scania has responded to this digital-era reality by building an international database to improve fleet management.[18] Scania can monitor how full a truck is and can assist clients to improve upon the 60 percent fill rate of trucks on European roads.[19] Scania has moved beyond truck manufacturing and now trades in transactional data about speed, fuel use, engine performance, and driving technique – information needed to reduce costs across the entire transportation market.[20]

We return to the importance of understanding how data streams can create value in the following chapter. Here, it is important to recognize that being a technology taker is not a strategy that lends itself to guesswork. Like Scania, organizations must objectively access and analyze continuous, real-time data for strategic planning purposes.

PLAN FOR A DIGITAL-ERA STRATEGY

A strategic plan encompasses the continuous review of an organization's vision, mission, strategic objectives, and guiding principles. Technology taking is a choice to join the digital era, instead of remaining in the enabling era's past. To do so, technology takers reflect their choices in strategic objectives and guiding principles.

Players in the enabling era were involved in a single game of change, stuck with a Tailor's set of internally defined parameters for success. Any changes occurred through a well-defined, beginning-and-end change transformation project. An enabling strategy relied on customized, internally defined processes to assist the efficiency and effectiveness of internal and external stakeholders.

In contrast, the digital era represents iterative technology-taking. The organization must constantly, as a matter of strategy, manage the adoption and adaptation of externally defined, continuously updating technologies. However, technology-taking and related change management is not a singular event. Planning for a digital-era strategy as a technology taker requires embarking on a continuous journey and not a singular project.

Recognizing that the digital era is still evolving, we observe that technology taking most often occurs among the Early Majority of the Innovation Adoption Lifecycle.[21] Adopting a new technology after Innovators (2.5%) and Early Adopters (13.5%), the Early Majority represents 34% of the potential audience.[22] Once leading industries and organizations have chosen to use this technology as a best practice and to take advantage of the particular technology's benefits, it is time for would-be technology takers to jump aboard. The risk does not come from adopting the new technology; it comes from staying behind and trying to use an outdated technology.[23] By the time most of the Early Majority is paying attention, a new technology is well on its way to becoming a dominant force in the market. If the market has moved ahead and established a new technology as a best practice, then not to adopt this technology is wasteful in terms of efficiency and ineffective in terms of the organization's mission.

The digital era is too great a leap beyond the enabling era for an organization to try to maintain a foot in each. Planning for each of the four areas of the Adoption—Adaptation Strategy Matrix is not a recommended approach. An organization cannot both customize and accept wholesale new technologies. Neither can the organization decide that its change management efforts have finished while continuing to use, say, a cloud-based ERP. And an organization cannot pretend that, while using digital-era technologies to become more efficient and effective, its strategy, let alone its strategic planning, can remain a discrete, rather than a continuous, exercise.

Like price takers who must accept the market they are in and adapt to its realities, the technology taker too must constantly react to maintain relevance for its stakeholders. Upgrades are not a choice; they are set by the technology maker. Technology takers must accept the realities of the technologies they have adopted and temper their reactions to them in order to further their mission. Launching and then upgrading an ERP at will is not an option with a cloud-based, SaaS solution. Apple slows down its smartphone batteries,[24] and Tesla remotely updates the braking distance of their Model 3,[25] regardless of users' desires or actions.

Consider the bank HSBC's maiden use of blockchain to finance the soybean trade between entities in Argentina and Malaysia.[26] The HSBC deal was the first use of a single, shared digital-era application for letters of credit and a network of computers to verify trade transactions.[27] The blockchain is not a technology that merely enabled a process by using a computer. Instead, blockchain is the quintessential digital-era technology, in that the underlying distributed-ledger verification process used cannot be altered by any one party. Users of the technology must modify their own behaviors and business practices to use it.

Whereas digitizing shipping documents would have brought large efficiency gains to shippers, prior to blockchain, no common infrastructure could support the use of electronic documents and coordination of multiple parties.[28] Blockchain technology reduced the risk of fraud in trade transactions and the time required to exchange trade information.[29] Adoption of the blockchain for trade finance of agricultural commodities is a potential seismic strategic change analogous to the adoption of standardized shipping containers by ships, ports, railways, and trade companies.[30]

Halting organizational evolution at the enabling era and not jumping to the digital era will limit the organization's ability to build required change into its strategic planning. HSBC and Cargill strategized to find a market edge by going beyond enabling to behavior-changing technologies. Each organization leaps to the digital future through planning that helps them execute technology-taking as a strategy. Recognizing the need to adapt to the digital era or see their business wither, HSBC and Cargill adopted the blockchain and big data.

Increased Interconnectivity

Behavior change and the ability to analyze data are prerequisites for strategy formulation in the digital era, as is understanding that cloud-based technologies have connected competing and noncompeting organizations. Technology takers should recognize the key to modern strategic planning is "the positioning of one business against others."[31] In the digital era, the same two businesses may now be technology takers of the same technologies. They compete head-on using exactly the same tech-enabled processes.

Amazon's fastest growing profitability segment in 2018 was fees from third-party vendors.[32] These vendors compete directly with Amazon to sell their products – on Amazon.com. In exchange, these vendors pay Amazon fees to use its globally accessible platform. The vendors also give Amazon their data, creating a symbiotic relationship between them.

Amazon supports a strategy of increasing revenues based on data stream analysis that third-party sales drive its profitability.[33] At the same time, over 100,000 businesses used Amazon's platform to profit, each generating more than US$100,000 in sales in 2017.[34]

Digital-era strategy accepts that an organization exists as part of a system composed of transactions between itself and its stakeholders, such as customers, employees, suppliers, and shareholders.[35] It does so because organizational interconnectivity within a system of transactions has increased dramatically due to technology-taking. The use of globally available processes, like the HSBC blockchain or the Amazon platform, connects multiple organizations' data. These digital-era processes align transactional data streams across organizations and their units.

With digital-era technologies, process-level interconnectivity has leapt beyond what was anticipated even a few years ago. The ERP customizations were to enable internally defined best practices. But the customizations left behind the organizations that demanded them, as the world rushed to use common processes available on cloud-based SaaS. Because digital-era technologies now drive the market, previously resistant organizations have been forced to become technology takers and not process customizers. Large, cloud-based technologies like SaaS now also allow direct benchmarking between interconnected organizations all the way down to the business process. Or, put into terms of cascading levels of strategy from corporate to business unit and functions, for technology takers, the level of interconnectivity links even the unit and functional levels of organizations.[36] This linkage will affect how organizations create competitive advantage in each of the businesses in which they compete.[37] Strategic planning without awareness of digital-era interconnectivity would be foolhardy, as the organization's profits and value are ultimately at risk; not doing so threatens the organization's mission.

Digital-Era Leadership Challenges (You Can't Lead What You Don't Understand)

Technology taking as a strategy will work only if implemented by leaders who are willing and able to formulate and plan strategies fit for the modern age. Most senior executives have a very hard time grasping the concepts of choosing the digital era, basing strategies on data, and factoring in interconnectivity. Many leaders have, at best, a rudimentary understanding of how enabling technologies have evolved to drive behavior change and to create value. Further, organizations' leaders and managers have difficulty envisioning technology as anything more than doing paper-based processes in an electronic way.

The US Congress' questioning of Mark Zuckerberg, the Facebook CEO, about the data breaches that occurred at his company illustrates the leadership challenge. The US representatives and senators revealed themselves to have a paucity of knowledge of how the Internet works and seemed to understand little about the digital-era technologies applied by Facebook.[38] Congressional leaders, seeking to lead in the regulation of the most successful tech-driven economy in the world, could not ask appropriate digital-era questions.

They are not alone. Organizations that are not themselves in the technology sector tend to have leadership that lacks a basic understanding of how the digital era affects their businesses. One level down from the leadership, most economists, medical doctors, lawyers, politicians, and manufacturing workers too have but a rudimentary understanding of technology. Tech is something that is delegated − traditionally, to the bottom of the Behavior Change Delta.

Yet each of the economic, medical, legal, political, and manufacturing sectors is being upended, changing rapidly through disintermediation and increased levels of interconnectivity via digital-era technologies. There are strategic ramifications if organizational leaders and managers are not technologically savvy. To fail to understand the digital era is to fail to plan for its effects and to fail to generate value from it. It is hard to see how organizations, let alone societies, can be successful in the digital era without strategically planning to take the leadership challenge of technology-taking head-on.

WHAT IS YOUR STRATEGY FOR THE DIGITAL ERA?

Are digital-era technologies causing dislocations in your industry? Do you know enough about these technologies to implement a Technology Taker Strategy; or, are you charting a wayward path as a Maker, Tinker, or Tailor? Are there data from which you can develop a fact-based, risk-managed strategy? Do you plan to manage change consistently and lead the adoption and adaptation needed to thrive in the digital era?

The strategy will need to cover all three of the cascading levels of organizational strategy: corporate, business unit, and function. Each has a role in how to operate in the digital era. The increase in interconnectivity and risk of disintermediation affect each level of strategy development. Your strategy also needs to capture goals, the organization's value proposition, growth ambitions, resource allocation model, and the role of senior management in executing strategy.

The underlying assumptions for corporate strategy are usually make or buy decisions regarding technology. The technology-taking strategic premise is not a buy or make decision in the traditional sense. The focus is instead on balancing the levels of adaptation and adoption of technologies that can best support strategic intent, which can be recognized as the link between the established vision and the current market situation of the organization. To link vision to the market in which you are operating, overarching, measurable, and time-limited goals need to be developed. For the technology taker, those goals must relate to the level of adaptation and adoption of digital era technologies that the organization is committed to achieving.

For each business unit, technology will drive behavioral change, but not necessarily to the detriment of people and existing processes in the market space in which the organization operates. Technology, process, and people aspects of strategic planning need to co-exist; and industry dynamics also matter. A Porter-inspired Five Forces approach to analyze the competitive nature of an industry is still valid for understanding and setting out the strategic positioning of both the corporation and each of its business units.[39]

At the lowest level of strategic planning, an organization's functional footprint may need to change to match its technology-taking organizational and unit strategy. Traditionally, functional strategy work has focused on specific descriptions of the tasks and related tactics of each business unit. With identifiable, measurable goals, functional descriptions will explain how units will operate and become more than the sum of their parts.

Where the organization is on the Adoption–Adaptation Strategy Matrix matters, but which direction the organization plans to go toward matters even more. The direction toward the Tailor, Tinker, Maker, or Taker quadrant is what must be captured in the organization's strategy. The strategy then can guide the development and execution of tactical steps for business units and functional levels.

The ability to adopt digital-era technologies may become its own force for the technology taker to confront. If there are no adoptable technologies available to one organization, then other organizations are not adopting them, either. If that is the case, then the strategic risk of disintermediation is reduced and that should have an effect on strategy. In this situation, there may be the possibility of innovating to become a technology maker. In most cases, however, a move north-east in the Adoption–Adaptation Strategy Matrix toward technology taking is the more likely strategy.

STRATEGY FOR THE FUTURE

To appreciate the advantages of the digital era, technology takers must be guided by the realization that the view ahead must be long and the goalposts constantly moving. Classic strategic plans typically are three-to-five-year strategies with action plans to deliver on their strategic objectives. For an organization to use technology-taking as a strategy, change leaders must support strategic planning and strategic objectives that seek digital era readiness. Planners need to set out time-limited goals for strategic objectives and to revise these goals continuously as they are met or the technology changes. And technology-taking change management plays should be used to meet the goals set by supporting readiness.

The change management playbook at Chapter 4 suggests exactly how strategic planning can be aligned with the modern era. A technology-taking strategy requires a close look at all organizational functions' ability to create value – now and in the ever-changing future. A new, high-level, and permanent change management function should support change leaders with the adoption of digital-era technologies and the adaptation of workplace behavior to these technologies. Plays related to envisioning, governing, engaging, investing, and measuring change will ensure upward and forward movement in the Adoption–Adaptation Strategy Matrix toward a more sustainable position vis-a-vis technology.

NOTES

1. But see Westerman, G., Bonnet, D., & McAfee, A. (2014). Leading digital: Turning technology into business transformation. *Harvard Business Review Press*, Massachusetts (outlining how a range of large firms, for example, finance, manufacturing, and pharmaceuticals, are using digital technologies for strategic advantage; and also outlining a number of practices and principles for successful digital transformation).

2. Cotula, L. (2013). The new enclosures? Polanyi, international investment law and the global land rush. *Third World Quarterly, 34*(9), 1605–1629. doi:10.1080/01436597.2013.843847

3. Polanyi, K. (1945). *Origins of our time: The great transformation.* London: V. Gollancz.

4. Zabecki, D. (May, 2015). *Military developments of world war I.* Retrieved from https://encyclopedia.1914-1918-online.net/article/military_developments_of_world_war_i

5. Meffert, J., & Swaminathan, A. (October, 2017). *Management's next frontier: Making the most of the ecosystem economy.* Retrieved from https://www.mckinsey.com/business-functions/digital-mckinsey/our-insights/managements-next-frontier

6. The last Kodak moment? (January 14, 2012). Retrieved from https://www.economist.com/node/21542796

7. Satell, G. (September 21, 2014). *A look back at why blockbuster really failed and why it didn't have to.* Retrieved from https://www.forbes.com/sites/gregsatell/2014/09/05/a-look-back-at-why-blockbuster-really-failed-and-why-it-didnt-have-to/#50ceb9431d64

8. Thompson, D. (May 17, 2018). *Disneyflix is coming. And Netflix should be scared.* Retrieved from https://www.theatlantic.com/magazine/archive/2018/05/disneyflix-netflix/556895/

9. Managing change: How law firms are answering the wake-up call, 35(5) *Law Practice* 32 (July/Aug 2009).

10. Leary, K. (February 27, 2018). *The verdict is in: AI outperforms human lawyers in reviewing legal documents.* Retrieved from https://futurism.com/ai-contracts-lawyers-lawgeex/

11. Järild, A. (n.d.). *How digital disintermediation is disrupting food and financial advice.* Retrieved from https://blog.thinque.com.au/how-digital-disintermediation-is-disrupting-food-and-financial-advice

12. Sinclair, J., & Wilken, R. (2009). Sleeping with the enemy: Disintermediation in internet advertising. *Media International Australia, 132*(1), 93–104. doi:10.1177/1329878x0913200110

13. Mintzberg, H. (February, 1994). *The fall and rise of strategic planning.* Retrieved from https://hbr.org/1994/01/the-fall-and-rise-of-strategic-planning

14. Marshall, A. (November 17, 2017). *Will Tesla's automated truck kill trucking jobs?* Retrieved from https://www.wired.com/story/what-does-teslas-truck-mean-for-truckers/

15. Siddiqui, F. (March 31, 2018). *Why D.C. is targeting the ride-hail industry.* Retrieved from https://www.washingtonpost.com/local/trafficandcommuting/why-dc-is-targeting-the-ride-hail-industry/2018/03/31/ef01fca8-3473-11e8-94fa-32d48460b955_story.html

16. Winsor, S. (April 16, 2015). *Adopt big data, or else.* Retrieved from http://www.cornandsoybeandigest.com/precision-ag/adopt-big-data-or-else

17. Ibid.

18. Beattie, A. (May 13, 2018). *Data protectionism: The growing menace to global business.* Retrieved from https://www.ft.com/content/6f0f41e4-47de-11e8-8ee8-cae73aab7ccb.

19. *Scania One – the digital platform for connected services.* (September 29, 2017). Retrieved from https://www.scania.com/group/en/scania-one-the-digital-platform-for-connected-services/

20. Beattie, A. (May 13, 2018). *Data protectionism: The growing menace to global business.* Retrieved from https://www.ft.com/content/6f0f41e4-47de-11e8-8ee8-cae73aab7ccb

21. Bohlen, J. M., & Beal, G. M. (May, 1957). *The diffusion process.* Special Report No. 18. Agriculture Extension Service, Iowa State College

22. Rogers, E. M. (2003). *Diffusion of innovations* (5th ed.). New York, NY: Free Press.

23. Anthony, S. (July 23, 2014). *First mover or fast follower?* Retrieved from https://hbr.org/2012/06/first-mover-or-fast-follower

24. *A message to our customers.* (December 28, 2017). Retrieved from https://www.apple.com/iphone-battery-and-performance/

25. Lambert, F. (May 31, 2018). *Tesla Model 3 stopping distance improvements confirmed in new test, Musk says UI/ride comfort improvements coming.* Retrieved from https://electrek.co/2018/05/30/tesla-model-3-stopping-distance-improvements-new-test-ui-ride-comfort-road-noise/

26. Chatterjee, S. (May 14, 2018). *HSBC says performs first trade finance deal using single blockchain system.* Retrieved from https://uk.reuters.com/article/uk-hsbc-blockchain/hsbc-says-performs-first-trade-finance-transaction-using-blockchain-idUKKCN1IF03H

27. Ibid.

28. Ibid.

29. Describing the South African Reserve Bank's test of blockchain technology to process 70,000 interbank wholesale settlement transactions in less than two hours: SARB's blockchain test: Typical daily SA interbank settlements done in under 2 hrs. (June 06, 2018). Retrieved from https://www.biznews.com/global-investing/2018/06/06/sarb-blockchain-pilot-daily-interbank-settlements/. See also Chatterjee, S. (May 14, 2018). *HSBC says performs first trade finance deal using single blockchain system.* Retrieved from https://uk.reuters.com/article/uk-hsbc-blockchain/hsbc-says-performs-first-trade-finance-transaction-using-blockchain-idUKKCN1IF03H

30. *Agriculture Blockchain Technology.* (n.d.). Retrieved from https://ccgrouppr.com/practical-applications-of-blockchain-technology/sectors/agriculture/

31. Kenny, G. (April 30, 2018). *Your strategic plans probably aren't strategic, or even plans.* Retrieved from https://hbr.org/2018/04/your-strategic-plans-probably-arent-strategic-or-even-plans

32. Levy, A. (April 27, 2018). *Amazon's sellers are going global, helping the company generate big profits.* Retrieved from https://www.cnbc.com/2018/04/26/amazon-25-percent-of-third-party-sales-came-from-global-sellers.html

33. Bezos, J. (n.d.). (1997). *Letter to shareholders.* Retrieved from https://www.sec.gov/Archives/edgar/data/1018724/000119312518121161/d456916dex991.htm

34. Ibid.

35. Strategic planning recognizing an organization's connection with other organizations also considers Porter's Five Forces or Goold's division of cascading corporate, unit, or functional levels. See Kenny, G. (2018, April 30). *Your strategic plans probably aren't strategic, or even plans.* Retrieved from https://hbr.org/2018/04/your-strategic-plans-probably-arent-strategic-or-even-plans; and see also Mind Content Tools Team. (n.d.). *Porter's five forces: Understanding Competitive forces to maximize profitability.* Retrieved from https://www.mindtools.com/pages/article/newTMC_08.htm

36. Campbell, A., Goold, M., & Alexander, M. (March/April, 1995). *Corporate strategy: The quest for parenting advantage.* Retrieved from https://hbr.org/1995/03/corporate-strategy-the-quest-for-parenting-advantage

37. Ibid.

38. *Confusing questions Congress asked Zuckerberg – CNN video.* (April 11, 2018). Retrieved from https://www.cnn.com/videos/cnnmoney/2018/04/11/facebook-zuckerberg-confusing-questions-congress-cnnmoney-orig.cnnmoney

39. Mind Content Tools Team. (n.d.). *Porter's five forces: Understanding competitive forces to maximize profitability.* Retrieved from https://www.mindtools.com/pages/article/newTMC_08.htm

CHAPTER 3

CREATE VALUE THROUGH DATA ANALYSIS AND BEHAVIOR CHANGE

When Tamara Mellon left her shoe company Jimmy Choo in 2011, she channeled her lessons learned into creating another luxury footwear business. Jimmy Choo's were (and still are) sold wholesale to high-end retailers throughout the world. But, based on Mellon's observation about the dominance of online shopping, her new, eponymous company sells "direct-to-consumer."[1] Tamara Mellon shoes are sold exclusively via the company's website. Data about each customer are tracked rigorously and in real time, allowing the company to react and change quickly.[2] The company can target each customer with particular ads, and the stock is adjusted based on immediate demand per location. Mellon says that she was challenged to learn a new way of operations and a new business language around "this tech-powered way of working."[3] But e-commerce is a reality of the digital era, so companies like Tamara Mellon must understand what the data are telling them and evolve to respond.

The investment community believes that Tamara Mellon is positioned to continue creating value. The company set out to raise US$15 million in Series B financing, and they closed the round at US$24 million.[4] By adopting an online-enabled, direct-to-consumer model, exploiting customer data, and taking rapid action, Tamara Mellon has created significant value in her brand.

WHAT IS VALUE?

"Value creation" has a different meaning for each industry and organization. An organization creates value for its stakeholders when it executes its mission. For example, a massive study of the service industry's use of data analysis found that an increase in value could be described as safer driving (for auto infotainment services), stable equipment operations (for providers of health prognostics), better fitness tracking (for exercise monitors), better heart condition monitoring (for post-operative providers of cardiovascular patients), and efficient energy management (for building managers).[5]

Nonprofit organizations and government agencies are no different from for-profit companies in this regard. By executing its mission, an organization creates value for its stakeholders, who may include donors, taxpayers, and constituents. Over 20 years ago, the City of Charlotte began using a Balanced Scorecard to track value creation. One of the objectives for the fiscal year 2017–2018 is to "foster economic success for everyone in the community." Progress toward this objective is measured in several ways, including the miles of new sidewalks and bikeways built per year. For the fiscal year 2017–2018, Charlotte's target is to build 10 miles of new sidewalks and bikeways to enhance safety, provide transportation choices, and better connect residents to employment opportunities and services.[6] While Charlotte does not have a direct measure of "economic success for everyone," the bottom line is that the city continues to attract new residents. Today, Charlotte is the third fastest growing city in the US.[7]

Nonprofits often look at value gained as a return on investment,[8] such as Charlotte's investment in sidewalks and bikeways. Private companies regard value gained as an increase in revenue and the company's overall profitability. Facebook, for example, estimates that each of its five billion users is worth US$82 in advertising revenue per year.[9]

Some organizations still attempt to justify implementing new technology by reducing costs. This might come in the form of streamlined processes or reduced workforce, or both. While this approach may be necessary for certain circumstances, it can pose an additional change leadership challenges. A better way is, where possible, to keep up to date, rather than make a sudden, large change that may then result in layoffs.

Being a technology taker means to create value – however, the organization defines it – through digital-era technologies and appreciating both their potential efficiencies and potential effectiveness. The technology taker does not gain value by customizing the technology. Instead, the technology-taking organization becomes more effective by analyzing the data collected to gain insight into their customers and markets, taking action and changing behaviors based on those insights, and achieving improved business outcomes.[10] Analysis of data generated by a digital-era system provides information to indicate which behaviors must be changed to achieve better the organization's mission.

While data analysis can occur at speed[11] lasting behavioral change takes months and sometimes years.[12] Successful technology takers build organizational agility and change capability (through use of the Playbook to Digital Era Change Leadership, at Chapter 4) so that behavioral changes can begin more quickly and be sustained more deeply.[13]

HOW TO GENERATE VALUE

Rather than spending a lot of time, energy, and money developing unique, tailored technology solutions, technology takers are freed to focus on fostering business value. Technology takers can create additional value in several ways, including reducing or avoiding costs, improving operations, increasing revenues or funds, or meeting customer needs and expectations.

First, the value can be found through cost-avoidance or reduction. By avoiding software customization, technology takers avoid long-term never-ending investment. As the modern world continues to evolve, a proprietary system would need to be designed and redesigned to react. Technology-taking, that is, adopting standard processes delivered by those in the business of technology making, seems the lesser cost than the continual investment in recustomization. Technology takers may also avoid costs by expanding operations and reaching new markets without having to hire additional employees. And they may reduce costs when certain tasks are eliminated or automated.

Second, with better and more timely data, and assuming they have the requisite skill and will, managers can make better decisions, leading to improved business operations. Value can be created through the replacement of less efficient processes with more efficient ones defined by globally dominant technologies. In this, the very act of being a technology taker results in increased value through increased efficiency in work practices. Walmart and other retailers are moving from brick-and-mortar stores to online retailers featuring free, overnight shipping.[14] These attempts, ostensibly take up the business practices pioneered by Amazon and to regain some of Amazon's market share, also recognize the revolution of shoppers' behavior in the digital era. Much shopping is now done via the Internet, so retailers are wise to focus their efforts where they find their consumers.

However, making processes more efficient or avoiding customization costs will not, in the long term, be the sole factors supporting value creation. Platforms to access and analyze this vast dataset offer organizations almost limitless ways to commoditize information. As a third way of creating value, organizations can increase revenue by creating new products and services that sell or control this information for increased efficiency and effectiveness. Electronic health records,[15] with their comparable and saleable data, and Internet cat videos,[16] for reasons than somewhat escape these authors, exemplify the creation of new, digital-era products and services.

Finally, the importance of automation goes beyond streamlining administrative and transactional tasks to taking advantage of faster, better service delivery, and enhancing the customer experience.[17] According to Accenture, 80% of brands believe they deliver great customer experiences; eight percent of consumers agree.[18] The challenge is to embrace the digital era in a way that creates value for the customer experience and enables employees to spend more of their time on higher value activities.

Value Through Cost Avoidance and Reduction: Universities in the Middle East

The adoption of a digital-era technology may lead to value creation. Cloud computing is a powerful technology to perform massive-scale and complex computing, which eliminates the need to maintain expensive computing hardware, dedicated space, and software.[19] Using cloud services allows organizations to own the minimum sufficient hardware to enable them to connect. The organization thus realizes a reduction in equipment costs and also can eliminate maintenance costs for these unneeded servers, networks, and security appliance hardware.[20] Software maintenance costs too are reduced, as are recovery costs from system failure and costs of managing both people and the system.[21] System upgrades are regular and at a consistent, predictable cost, permitting the organization to plan better for them. Capital expenditures are also decreased or eliminated because they no longer are required. As the organization needs change, it can stop using and stop paying for whatever cloud-based application that is no longer needed.[22]

A review of Middle Eastern universities' cloud computing adoption found that these digital-era systems could reduce the size of information technology departments from 10 to 15 employees to three employees. People no longer are required to revise the systems manually because cloud-based systems update continuously. Further, staff do not need to monitor software licenses and updates, as revisions occur in the cloud:[23] the entire data center would be updated every three to five years. Going to the cloud made all these data center costs disappear into the ether.

Yet an organization cannot rely on the mere adoption of a digital-era technology to generate value. For all the costs saved by reducing hardware, software, maintenance, and people, additional costs will be incurred through hiring people with new, required skills. Digital-era systems are disruptive of existing work and ways of doing things, but stopping existing projects and turning to new ones also involve costs of staff time and training.[24]

The modern age requires adaptation and active engagement to generate value. Organizations must invest in developing their human and computing resources so that they can analyze the reams of data generated by new tech. In addition, organizations need to have in place management systems to change behaviors as indicated by data analysis.

Value Capture Through Data Analysis: Workday and Tesco

Digital-era technologies aggregate data acquired from multiple users. Compilation of data provides a rich information stream from organizations with the ability, skill, and resources to mine it. The joint data stream allows organizations to access information that would be impossible or too costly for one business alone to generate.[25] Value is created through an organizational use of data as a service to benchmark itself against others in the industry, to see how its practices and workers compare to or exceed norms and where efficiencies might be found. An additional source of value is through collaboration with other organizations using the same technology to maximize the capabilities of the technologies. Collaboration replaces the value-sapping practice of building security features to restrict competitors' knowledge about an organization.[26]

For example, the Workday Software as a Service (SaaS) Enterprise Resource Planning (ERP) encourages collaboration among its users, whose data the system collects, anonymizes, and agglomerates. Workday develops benchmarks from over 26 million users of the system's human resource management platform about employee turnover, workforce composition, and management effectiveness.[27] Each Workday client can determine how it compares to other organizations of similar size and industries. Workday users can liaise in the system's peer-to-peer interactive space, workday community, where the ERP's customers can share knowledge and experiences about using the technology. Workday monitors the community group and determines product enhancements based on the clamor for these in community.[28]

Tesco, the grocery giant, also uses real-time data analytics, Big Data, and the efficiencies made possible by the emerging Internet of Things to connect with a UK-wide network of stores and a worldwide network of distributors.[29] Tesco's market domination in the UK is made possible by technology taking of low-cost, open-source, "big-data" warehousing technology,[30] such as Teradata and Hadoop. Tesco uses these frameworks to analyze data sets about customer purchase patterns, product stocks, and company performance. Tesco was able to exploit this data to hold off competition from discount chains like Aldi and Lidl and to face down new threats from Amazon's grocery delivery service.[31]

Tesco created its data stream by innovating with the Tesco Clubcard, which the company introduced in 1995, far before its competitors.[32] Tesco gave its shoppers small discounts to induce use of their Clubcard whenever they shopped. By tracking each shopper's purchases, Tesco was able to see what products were bought when and where. The Clubcard served as a full-scale "change management instrument," revolutionizing the retail grocery business.[33]

Other grocers would quickly follow suit. The rate of value creation was disrupted for grocery chains that stayed behind, tinkering with their own existing, nondata generating sales programs. As the first grocer in the UK to combine and analyze shopper data and benchmark its own performance to that of its competitors, Tesco used technology-taking to strategic advantage. Tesco did not create the technologies it takes, but, through change management, Tesco has mastered technology-taking to meet its mission of "serving Britain's shoppers a little better every day."[34]

Creation of New Value Streams: Electronic Health Records and Cat Videos

The digital era's vast compilation of data has created new, never-before foreseen opportunities to create value. These highly lucrative value streams can range from the cutting edge, relating healthcare records to a patient's immediate care needs, to the extraordinarily mundane, driving ad revenue from the number of hits a cat video receives.

Using large, linked administrative databases to store and provide access to health data captured through electronic health records (EHR) allows for comparing the effectiveness of one therapy, say for cancer or Alzheimer's disease, versus another. Associating EHR-derived data with patient-level information can help doctors generate hypotheses about care, compare assessments, and offer personalized care. As implementation of EHRs increases, cloud-based databases containing information collected via EHRs will continuously update; aggregating data enhanced with real-time analytics can provide point-of-care evidence to doctors, tailored to patient-level characteristics.[35]

These healthcare data are a potential gold mine for companies able to exploit them. In early 2018, Facebook launched a project to induce hospitals to share anonymized patient data with the company. Facebook's plan was to combine these data streams with Facebook's own data trove of hospitals' patient data on diagnoses and prescription information. Facebook would then construct digital profiles of patients, even without these patients' consent. This information could be sold to pharmaceutical or managed care companies so that these companies could target advertisements to Facebook users potentially in need of certain medicines or treatments.[36] Direct-to-consumer internet pharma advertising, the very same underlying technology to value creation used by Tamara Mellon to sell shoes, was a US$4 billion industry in 2008. Since then, it has "mushroomed" in cost and scope.[37]

Whereas one might have predicted the creation of new value streams based on analysis of healthcare or shoe sales data, few could have anticipated the technology taking that has accompanied posting cat videos to the internet. There are more than two million cat videos on YouTube, with a total of about 26 billion views. That's an average of 12,000 views for each cat video, which is more views-per-video than any other category of YouTube content.[38] For the creators of the most-viewed cat videos, this is a big business.

YouTube was founded in 2005 to permit users to upload to the internet video content for anyone to see. Google now owns YouTube. Video creators can agree to let Google sell advertising that will appear on the creator's YouTube site, and the creator will receive a share of the ad revenue. The number of views of a particular video indicates audience size for ads. In 2014, the average rate for 30-second commercials that a viewer must watch before seeing a YouTube video was US$7.60 per 1,000 ad views.[39] (This pay rate has likely decreased since then, given that the number of YouTube videos has increased. Further, YouTube takes 45% of the total.) YouTube video like Nyan Cat, a flying cat with a rainbow trail set to synth music, was viewed 54 million times in 2011.[40]

Even more successful was the I Can Haz Cheezburger cat, a photo of an obese, hungry cat that seemed to be requesting a cheeseburger. The Cheezburger cat's picture attracted so many readers to the blog where it was initially posted that the creators, who were computer programmers, developed a dedicated Cheezburger cat website. Readers could send in their own cat pictures and other readers could vote on the submissions, awarding them from one to five cheeseburgers. To increase the site's viewers, the creators analyzed viewership data and timed posts for when people would most likely be reading.[41] By May 2007, five months after the website launched, it received 1.5 million views with up to 200,000 unique visitors per day. Touting its number of viewers, the Cheezburger cat site sold multiple ads costing between US$500 and US$4,000 per week. And by September 2007, the site's two creators sold the entire enterprise to investors for US$2 million.[42]

Cost Savings: United Nations and US Government Organizations

Despite the tantalizing possibility that anyone could spin gold from the straw of a funny cat image, most organizations are unable to develop an entirely new value stream from data collection and analysis derived from a digital-era technology. Organizations adopting digital-era technologies often must seek value by using modern tech to cut costs.

Cost savings, or the replacement of higher costs with lower costs, is a key value metric for many organizations' ultimate acceptance of digital-era solutions. Some nonprofit, government, and international organizations do not have performance metrics that link objective achievement to mission-based outcomes. Ideally, these organizations could show that their ascension to the digital era has driven their more effective achievement of their missions. Since successful missions to support health, strive for human rights, or serve the citizenry are difficult to measure precisely or to attribute to a single new technological adoption, their stakeholders may push the organization to demonstrate that an investment in a new technology will directly result in decreased costs.

Cost savings was exactly what was demanded of a UN organization that adopted a cloud-based ERP system at great cost and effort.[43] Because the new system was a digital-era one, significant behavior change was required of all people. Old procedures were abolished and new ERP-based ones established. This led to "anxiety," with staff demanding to know what had been achieved for all the work and money expended.[44] The organization's administration too was conscious that continued budget requests for funds allocated to the ERP system would be scrutinized by member governments. The ERP system needed quickly to prove its worth.[45]

Fortunately, mining the ERP's data stream provided answers to the question of cost-savings. By reviewing the ERP's reports, the UN organization was able to see which workers were spending the most time on which processes. In certain departments, cuts were made to staff that previously spent all working hours processing financial and human resources transactions; the new ERP business processes had replaced these people.[46] Additional reductions were made in the IT department, where there was no longer a need for people to develop, test, and deploy systems unique to the organization.[47] In the unit that responded to disasters and for the organization's business continuity plan, remarkable cost reduction was anticipated, as cloud service is designed with high redundancy and availability, making it reliable and secure.[48] (In fact, a study in disaster recovery found an 85% reduction in costs from the adoption of cloud-based computing.[49])

Yet the data about technology taking and cost savings are far from uni-directional, as the government of the United Kingdom has found. The UK

Government Digital Service (GDS) is a cabinet-level office that has led the country's adoption of innovative digital services. The GDS has spurred such rapid change that, three years after the GDS' founding in 2011, the UK's reputation moved from "a wasteland of IT failures" to one of "the most digitally advanced governments in the world."[50]

Notwithstanding the GDS' "potent, government-wide powers," the vaunted talent of its people, and the digital-era policy changes the unit has driven across the English-speaking world, the value of GDS-spurred technology taking cannot be described uniformly in cost savings terms.[51] GDS is praised as having produced new or improved government services in less time and at higher quality. Of the 25 "exemplar" services GDS prioritized as part of its initial work program, 12 of these will see benefits outweigh costs of development within 10 years.[52] But 10 of those services will still see development costs outweigh expected benefits in the same time period.[53] Value creation must be found elsewhere than through cutting costs.

RESULTING BEHAVIOR CHANGES (WHAT ARE THE DATA TELLING US?)

Creating value through technology taking does not depend only on mining the data generated by digital-era technologies. There is a purpose to the data analytics exercise: to foment behavior change. Tamara Mellon did not solely observe that consumers now buy shoes online and want the latest fashions available immediately to them. She also developed a business model based on internet advertising directed to the individual shoe buyer and targeting that buyer's interests, previous purchases, requests, and views of products. The digital era changed the shoe market, and Tamara Mellon changed her behavior to meet the new market's demands.

Behavior change through change management of technology-taking may be the preeminent driver of value creation in the digital era. While digital-era technologies may improve organizing at systems and operations levels, the development of unique competencies of skills and knowledge acquired through the implementation of digital era technology is what is likely to lead to the creation and capture of value relevant for competitiveness.[54]

Data Driving Behavior Change: Lyft and Uber

The ride-hailing and sharing businesses Lyft and Uber are the very embodiment of the behavioral responses driven by digital-era technologies.[55] Lyft and Uber each use a proprietary technology downloaded as a smartphone application by potential riders and drivers. This app matches ride demand with driver supply. The technology also rationalizes the supply of vehicles, directing drivers to locations where they are requested and limiting the total quantity of drivers in any one area. This supply rationalization occasionally results in a higher price for rides. But riders accept this higher price as a cost of being able to find a driver at times and in locations where a taxi might otherwise be unavailable.[56] The driver, who has accepted the costs of being an independent contractor, pockets 75% of the fare price; the ride-hailing company typically receives the other 25%.[57]

Drivers and riders are able to take advantage of the data generated by the technologies powering Uber and Lyft. These data are portable, which allows users to remotely access data or transfer these data to platforms, tasks, and institutional settings.[58] Portability is how Uber has become a verb. Trusting that a ride-share driver eventually will respond to a button tapped on an iPhone, riders now "Uber" their way to any possible destination. No more do many riders wait by the roadside, hoping a traditional taxi might pass their way. Confident that the company's technology will direct riders to them, drivers turn on their phones and wait for a rider to be added to their queue. Taxi drivers who do not drive for Uber or Lyft are experiencing an ever-decreasing market share, as the digital era has moved on without them.[59]

Uber and Lyft drivers have shown themselves as willing and able use a digital-era technology and to react as the tech requires. Similarly, in order to create value from big data generated by new tech, organizations must have the human and systems resources to synthesize those data and to derive meaning from them.[60] However, many organizations cannot appreciate the growth of value through technology taking because these businesses lack workers skilled in predictive modeling or the tools to disaggregate data and to align its information with processes (a topic to which we return in the Playbook to Digital Era Change Leadership at Chapter 4).[61]

Organizational Structure and People Permit
Change and Drive Value

In the digital era, tech cannot easily be used to replace workers. Too much reliance on algorithms for data analysis may lead to a loss or replacement of human knowledge, particularly when it is not clear how algorithms arrive at certain results, patterns, and decisions.[62] Digital-era technologies can provide a data stream and assist with its synthesis, but an organization's people remain responsible for analyzing data's meaning and using it to affect workplace practices.

Organizations must access, track, collect, manage, govern, process, and analyze data for data-driven decision making and implementation purposes. To develop such capabilities, organizations must develop, mobilize, and use technical and human resources. Once acquired, these technical and human resources must be integrated into the organizational structure.[63]

To reap the benefits of new technologies, an organization must have analysts and decision makers who can correlate and combine data in new ways to arrive at insights by exploring connections.[64] For data analysis to drive behavior change, organizations require creative and intelligent workers able to collaborate, instead of using the same pre-digital processes.[65] Digital-era workers further must be ethical, cognizant of the risks of data collection and analysis. Correlating different data sources (such as health and financial records – or even data about Uber rides to lovers or doctors[66]) may yield identification of private, sensitive information.[67]

The technology-taking organization's structure must permit working across organizational boundaries and not in silos. Organizational models should be flexible to facilitate cross-disciplinary interaction at the work level. Analytical teams should include different roles and perspectives to arrive at new, yet valuable insights.[68]

Limits on Behavior Change

No matter the data, the potential value, and the supportive organizational structure, people will change their behavior when they are ready. A prerequisite to "readiness" for behavior change is people understanding, believing, and accepting a personal reason to change.[69] If those who are making sense of data are biased by fixed, extant mindsets, their current daily routines, and historical values and norms, change will not occur.[70]

In general, most people may see no benefit in changing. While it is helpful if they see a benefit from the change, the perceived discomfort caused by change makes changes seem unnecessary. People must understand the costs and risks of not changing. For this reason, a playbook for digital era change management is required that, among other things, detail the responsibilities of change leaders for setting out a clear change vision, driving a business case in its support and implements change management to make sure that the costs of no change are communicated at a personal, not only organizational, level. Even so, once someone understands intellectually why she cannot continue the way she is, that person may still not be ready to change. Engagement by change leaders is required and even so, there may be a lacking ability to change. People may not have been fully equipped with the required skills. In those cases, the organization must invest in adequate training. Or they may lack confidence in the skills they have. Then, the organization's leader's vision for change needs to encourage them.

Because people will change their behavior only when it is in their interest, organizations must develop incentives for employees and managers to invest in their people. While challenging for most people, the embrace of technology will also need to be measured, as without it the organization may be left behind by technology takers who do welcome change.

Organizations in the digital era revolve around mobility and decreased loyalty and commitment. The modern age has given lie to previous theories assuming perfectly fungible human resources.[71] Investment in technology and in change management must be seen as a co-production generated by both the organization and its employees and for the betterment of both.[72]

Value creation is therefore rooted not exclusively in data analysis but also change management, especially in people's motivation to change.[73] Optimal value creation depends on an organization's ability to focus its people on common, joint goals for organizational success.[74] With technology-taking via tech adoption and behavior adaptation, people can begin to feel at sea, as though there is no constant anchor. In this environment, a shared sense of purpose, perhaps expressed in the organization's mission, transcends profits, and provides the ship a rudder.[75] Leaders are responsible for creating a sense of predictability through an unchanging purpose so that employees become willing to continually adapt to new technologies.

Organizations can become technology takers when their people become technology takers. To put it succinctly, the technology taker must be Tamara Mellon Inc. and Tamara Mellon. Both organization and worker must believe in the digital era's potential benefits for real value to be obtained from the changes the digital era requires.

NOTES

1. Draznin, H. (June 30, 2017). *Jimmy Choo co-founder: 'Society is better off when women earn equal' Jimmy Choo Co-Founder: "Society is better off when women earn equal."* Retrieved from http://money.cnn.com/2017/06/30/smallbusiness/tamara-mellon-jimmy-choo/index.html

2. Dunn, L. E. (June 05, 2017). *Women in business Q&A: Tamara Mellon.* Retrieved from https://www.huffingtonpost.com/entry/women-in-business-qa-tamara-mellon_us_59357964e4b0f33414194bf4

3. Ibid.

4. Segran, E. (June 05, 2018). *Luxury shoe startup Tamara Mellon just snagged $24 million.* Retrieved from https://www.fastcompany.com/40581360/luxury-shoe-startup-tamara-mellon-just-snagged-24-million

5. Lim, C. (2018). From data to value: A nine-factor framework for data-based value creation in information-intensive services. *International Journal of Information Management*, 121–135.

6. Charlotte Center City. (n.d.). *Center City 2020 vision plan.* Retrieved from https://www.charlottecentercity.org/center-city-initiatives-2/plans/2020-vision-plan/

7. Charlotte, North Carolina Population 2018. (June 12, 2018). Retrieved from http://worldpopulationreview.com/us-cities/charlotte-population/

8. Niven, P. R. (2010). *Balanced scorecard step-by-step: Maximizing performance and maintaining results.* Hoboken: Wiley.

9. Fowler, G. A. (April 05, 2018). *What if we paid for Facebook — instead of letting it spy on us for free?* Retrieved from https://www.washingtonpost.com/news/the-switch/wp/2018/04/05/what-if-we-paid-for-facebook-instead-of-letting-it-spy-on-us-for-free/

10. Haendly, M. (April 26, 2016). *5 tangible benefits of digital transformation.* Retrieved from https://sapinsider.wispubs.com/Assets/Articles/2016/April/SPI-5-Tangible-Benefits-of-Digital-Transformation

11. Boston Consulting Group. (n.d.). *Digital transformation — strategy for digitizing the business.* Retrieved from https://www.bcg.com/capabilities/technology-digital/digital.aspx

12. Mastrangelo, P. M., Prochaska, J., & Prochaska, J. (2008). How people change: The transtheoretical model of behavior change. *PsycEXTRA Dataset.* doi:10.1037/e518442013-832.

13. Subbiah, K., & Buono, A. F. (2013). "Internal Consultants as Change Agents: Roles, Responsibilities and Organizational Change Capacity". *Academy of Management Proceedings, 2013*(1), 10721. doi:10.5465/ambpp.2013.10721abstract.

14. Yohn, D. L. (July 25, 2017). *Walmart won't stay on top if its strategy is "Copy Amazon".* Retrieved from https://hbr.org/2017/03/walmart-wont-stay-on-top-if-its-strategy-is-copy-amazon

15. Greene, J. A., & Kesselheim, A. S. (2010). Pharmaceutical marketing and the new social media. New *England Journal of Medicine, 363*(22), 2087–2089. doi:10.1056/nejmp1004986

16. Tozzi, J. (July 13, 2007). *Bloggers bring in the big bucks.* Retrieved from https://web.archive.org/web/20080215230339/http://www.businessweek.com/smallbiz/content/jul2007/sb20070713_202390.htm.

17. *Top 10 digital transformation trends for 2018, Free Appian eBook.* (n.d.). Retrieved from https://sf.tradepub.com/free/w_appf228/

18. *Creating the best customer experience | Accenture Interactive.* (n.d.). Retrieved from https://www.accenture.com/us-en/interactive-index

19. Yang, C., Huang, Q., Li, Z., Liu, K., & Hu, F. (2017). Big data and cloud computing: innovation opportunities and challenges. *International*

Journal of Digital Earth, 10(1), 13–53. doi:10.1080/
17538947.2016.1239771

20. Al-Badi, A., Tarhini, A., & Al-Kaaf, W. (2017). Financial incentives
for adopting cloud computing in higher educational institutions. *Asian
Social Science, 13*(4), 162. doi:10.5539/ass.v13n4p162

21. Lapouchnian, A. (June 01, 2011). *Exploiting requirements variability
for software customization and adaptation.* Retrieved from https://tspace.
library.utoronto.ca/handle/1807/27586

22. Al-Badi, A., Tarhini, A., & Al-Kaaf, W. (2017). Financial incentives
for adopting cloud computing in higher educational institutions. *Asian
Social Science, 13*(4), 162. doi:10.5539/ass.v13n4p162

23. Ibid.

24. Barreau, D. (2001). The hidden costs of implementing and maintaining information systems. *The Bottom Line, 14*(4), 207–213. doi:10.1108/
08880450110408481

25. Wu, F., & Cavusgil, S. T. (2006). Organizational learning, commitment, and joint value creation in interfirm relationships. *Journal of
Business Research, 59*(1), 81–89. doi:10.1016/j.jbusres.2005.03.005

26. Ibid.

27. *How workday is doubling down on data and analytics.* (October 12,
2017). Retrieved from http://blogs.workday.com/how-workday-is-dou-
bling-down-on-data-and-analytics/

28. *Workday community.* (n.d.). Retrieved from https://www.workday.
com/en-us/company/about-workday/community.html

29. Marr, B. (November 17, 2016). *Big data at Tesco: Real time analytics
at the UK grocery retail giant.* Retrieved from https://www.forbes.com/
sites/bernardmarr/2016/11/17/big-data-at-tesco-real-time-analytics-at-the-
uk-grocery-retail-giant/3/#1d6afed51333

30. Oracle Big Data. (n.d.). Retrieved from https://www.oracle.com/big-
data/guide/what-is-big-data.html

31. Marr, B. (November 17, 2016). *Big data at Tesco: real time analytics
at the UK grocery retail giant.* Retrieved from https://www.forbes.com/

sites/bernardmarr/2016/11/17/big-data-at-tesco-real-time-analytics-at-the-uk-grocery-retail-giant/3/#1d6afed51333

32. Describing a paper by Simon Knox, Cranfield University School of Management, Tesco, Building a Global Retail Brand Through Sustainable Marketing): Tesco: A measurable marketing case study. (July 25, 2012). Retrieved from https://www.smartcompany.com.au/people-human-resources/managing/tesco-a-measurable-marketing-case-study

33. Ibid.

34. Tesco PLC. (n.d.). *Core purpose and values.* Retrieved from https://www.tescoplc.com/about-us/core-purpose-and-values/

35. Miriovsky, B. J., Shulman, L. N., & Abernethy, A. P. (2012). Importance of health information technology, electronic health records, and continuously aggregating data to comparative effectiveness research and learning health care. *Journal of Clinical Oncology, 30*(34), 4243–4248. doi:10.1200/jco.2012.42.8011

36. Ostherr, K. (April 18, 2018). *Perspective | Facebook knows a ton about your health.* Now they want to make money off it. Retrieved from https://www.washingtonpost.com/news/posteverything/wp/2018/04/18/facebook-knows-a-ton-about-your-health-now-they-want-to-make-money-off-it/

37. Greene, J. A., & Kesselheim, A. S. (2010). Pharmaceutical marketing and the new social media. New *England Journal of Medicine, 363*(22), 2087–2089. doi:10.1056/nejmp1004986

38. Myrick, J. G. (2015). Emotion regulation, procrastination, and watching cat videos online: Who watches internet cats, why, and to what effect? *Computers in Human Behavior, 52,* 168. doi:10.1016/j.chb.2015.06.001

39. Kaufman, L. (February 1, 2014). *Chasing their star, on YouTube.* Retrieved from https://www.nytimes.com/2014/02/02/business/chasing-their-star-on-youtube.html

40. Netburn, D. (December 20, 2011). *Talking twin babies, Nyan Cat among YouTube's top videos of 2011.* Retrieved from http://latimesblogs.latimes.com/technology/2011/12/talking-twin-babies-nyan-cat-and-friday-top-youtubes-most-watched-videos-of-2011.html

41. Tozzi, J. (July 13, 2007). *Bloggers bring in the big bucks.* Retrieved from https://web.archive.org/web/20080215230339/http://www.businessweek.com/smallbiz/content/jul2007/sb20070713_202390.htm

42. Cox, T. (October 21, 2008). *The kitty site that's a phenomenon.* Retrieved from https://www.thetimes.co.uk/article/the-kitty-site-thats-a-phenomenon-5nh9bfpskzg

43. *A cloud-based ERP renovates work practices and changes behavior at PAHO* (Case Study Series, pp. 1–15.). United Nations System Staff College. Retrieved from http://www.unssc.org/sites/unssc.org/files/mini_case_study_unssc_02_fin.pdf

44. Ibid.

45. Ibid.

46. Ibid.

47. Ibid.

48. Clarke, A. (2017). Digital government units: Origins, orthodoxy and critical considerations for public management theory and practice. *SSRN Electronic Journal.* doi:10.2139/ssrn.3001188

49. Al-Badi, A., Tarhini, A., & Al-Kaaf, W. (2017). Financial incentives for adopting cloud computing in higher educational institutions. *Asian Social Science, 13*(4), 162. doi:10.5539/ass.v13n4p162

50. Clarke, A. (2017). Digital government units: Origins, orthodoxy and critical considerations for public management theory and practice. *SSRN Electronic Journal.* doi:10.2139/ssrn.3001188

51. Ibid.

52. Ibid.

53. Ibid.

54. Mitra, A., Oregan, N., & Sarpong, D. (2018). Cloud resource adaptation: A resource based perspective on value creation for corporate growth. *Technological Forecasting and Social Change, 130*, 28–38. doi:10.1016/j.techfore.2017.08.012

55. Günther, W. A., Mehrizi, M. H., Huysman, M., & Feldberg, F. (2017). Debating big data: A literature review on realizing value from big data. *The Journal of Strategic Information Systems, 26*(3), 191–209. doi:10.1016/j.jsis.2017.07.003

56. Pullen, J. P. (November 04, 2014). *Everything you need to know about Uber*. Retrieved from http://time.com/3556741/uber/

57. Glon, R. (October 22, 2017). *How does Uber work? Here's how the app lets you ride, drive, or both*. Retrieved from https://www.digitaltrends.com/cars/how-does-uber-work/

58. Günther, W. A., Mehrizi, M. H., Huysman, M., & Feldberg, F. (2017). Debating big data: A literature review on realizing value from big data. *The Journal of Strategic Information Systems, 26*(3), 191–209. doi:10.1016/j.jsis.2017.07.003

59. Nelson, L. J. (April 14, 2016). *Uber and Lyft have devastated L.A.'s taxi industry, city records show*. Retrieved from http://www.latimes.com/local/lanow/la-me-ln-uber-lyft-taxis-la-20160413-story.html

60. Günther, W. A., Mehrizi, M. H., Huysman, M., & Feldberg, F. (2017). Debating big data: A literature review on realizing value from big data. *The Journal of Strategic Information Systems, 26*(3), 191–209. doi:10.1016/j.jsis.2017.07.003

61. Wang, Y., Kung, L., & Byrd, T. A. (2018). Big data analytics: Understanding its capabilities and potential benefits for healthcare organizations. *Technological Forecasting and Social Change, 126*, 3–13. doi:10.1016/j.techfore.2015.12.019

62. Günther, W. A., Mehrizi, M. H., Huysman, M., & Feldberg, F. (2017). Debating big data: A literature review on realizing value from big data. *The Journal of Strategic Information Systems, 26*(3), 191–209. doi:10.1016/j.jsis.2017.07.003

63. Ibid.

64. Ibid.

65. Ibid.

66. Perry, D. (November 20, 2014). *Sex and Uber's 'Rides of Glory': The company tracks your one-night stands – and much more*. Retrieved from

http://www.oregonlive.com/today/index.ssf/2014/11/sex_the_single_girl_and_ubers.html

67. Günther, W. A., Mehrizi, M. H., Huysman, M., & Feldberg, F. (2017). Debating big data: A literature review on realizing value from big data. *The Journal of Strategic Information Systems, 26*(3), 191–209. doi:10.1016/j.jsis.2017.07.003

68. Ibid.

69. Shah, N., Irani, Z., & Sharif, A. M. (2017). Big data in an HR context: Exploring organizational change readiness, employee attitudes and behaviors. *Journal of Business Research*, 70, 366–378. doi:10.1016/j.jbusres.2016.08.010

70. Günther, W. A., Mehrizi, M. H., Huysman, M., & Feldberg, F. (2017). Debating big data: A literature review on realizing value from big data. *The Journal of Strategic Information Systems, 26*(3), 191–209. doi:10.1016/j.jsis.2017.07.003

71. Mahoney, J. T., & Kor, Y. Y. (2015). Advancing the human capital perspective on value creation by joining capabilities and governance approaches. *Academy of Management Perspectives*, 29(3), 296–308. doi:10.5465/amp.2014.0151

72. Ibid.

73. Foss, N. J., & Lindenberg, S. (2013). Microfoundations for strategy: A goal-framing perspective on the drivers of value creation. *Academy of Management Perspectives, 27*(2), 85–102. doi:10.5465/amp.2012.0103

74. Ibid.

75. Birkinshaw, J., Foss, N. J., & Lindenberg, S. (n.d.). *Combining purpose with profits*. Retrieved from https://sloanreview.mit.edu/article/combining-purpose-with-profits/

CHAPTER 4

PLAYBOOK TO DIGITAL-ERA CHANGE LEADERSHIP

When equipped with a playbook for the digital era, leaders at any organization can support technology-driven behavior changes, including value creation through data analysis.[1] This playbook prescribes actions to be taken at all levels of an organization to harness technologically induced behaviors and to achieve organizations' goals. Written to guide technology takers through the disruption of the digital era, our playbook distinguishes itself from previous change management efforts.[2]

Today's digital technologies are driving continuous disruption. Most change management practitioners employ models of the past that apply to a well-defined, bounded change, that is, a project with a beginning, middle, and end. They often reference various three-phase models from change management practitioners such as Kurt Lewin (unfreeze, change, refreeze),[3] William Bridges (ending, neutral zone, new beginning),[4] Daryl Conner (present state, transition state, desired state),[5] and John Kotter (creating the climate for change, engaging and enabling the organization, implementing and sustaining the change).[6]

These three phases approaches do not work in the digital era, let alone for a technology-taking strategy. There is no third phase. Organizations are continually living in the middle phase, where everything is ambiguous and up in the air. Change can no longer be treated as a discrete, one-time event, even if classic change management efforts may still have their place for projects delimited with a clear beginning and end date. Single change management projects may also help an organization set new objectives and create sustainable value where an organization must make a hard stop and a quick turn-around by a certain date.[7]

But technology taking as a strategy is perpetual. Even moving from change projects to change project waves will not necessarily yield expected or successful results.[8] Technology taking has a very different structure and payoffs than a one-off change game. Digital-era technologies change constantly, and these changes cause still other changes. There are no invisible scissors to cut away all but a single change and to examine uniquely the effects of a particular project or effort. Instead, the ground is shifting under our feet.

CONTINUOUS CHANGE REQUIRES A NEW FORM OF CHANGE LEADERSHIP

Envisioning how to play in the digital era is a highly dynamic endeavor. Organizations need to build and maintain an organizational climate that supports and equips people to adapt to a continually changing landscape, to build personal resilience, to embrace innovation and transformation, and to thrive in a volatile environment. The digital era favors organizations that consciously choose to transform themselves. Leaders already recognize change resistance, but do not always recognize their limited ability to predict the future. Modern and effective leaders take risks by seeking technology-taking opportunities that arise from the chaos.

Technology-taking change leaders, those senior managers who direct technology taking as a strategy, are challenged to address never-ending technology changes while achieving the mission of their organizations. Leaders must shift from managing individual changes to managing continuous changes to building organizational and individual capacity for change. Often, they do so under protest. Digital-era technologies are metamorphosing ways of working, and keeping up with those changes is imperative for survival. At the same time, however, there is an ongoing business to attend to. Increased change capacity and resiliency can enable individuals and organizations to thrive in an environment of disruptive technologies, confusing markets, and organizational ambiguity.[9]

As organizations shift to accommodating continuous change, leaders are faced with skeptical employees accustomed to short-term change projects. For example, most classic leadership transformation initiatives take the form of a singular change event or project.[10] The knee-jerk reaction of most managers and employees is to "sit it out" until the initial energy of new leaders or the repercussion of a change project is better known. Efforts to change in those cases fail because "too many managers don't realize transformation is a process, not an event."[11]

Many want a momentary chance to re-group and relax. To extend Lewin's model, they would like to unfreeze, change, refreeze, and breathe. Unfortunately, some companies who decided to step aside and relax while their market changed around them have become irrelevant. Digital-era, continuous change management is a marathon, and organizations who attempt to sprint through it can die as employee exhaustion sets in.

Change projects that occur one after another are likely to overwhelm individual's capacity to adapt. Often new leaders' intention is to get "quick wins" or organizational culture change from a change event. Both instant change and the expectation of cultural transformation through a change management project are highly unlikely. Worse, these are usually a distraction, adding to exhaustion and resistance.

CHANGING ORGANIZATIONAL CULTURE

Cultural change is often the holy grail of change efforts. Organizational culture shapes attitudes and behaviors through shared assumptions and behavioral norms. These norms determine which behaviors are encouraged, discouraged, accepted, or rejected within an organization.[12] But, anthropologically speaking, culture is always changing.[13] An intention of managed culture change can create false expectations that no change manager can achieve.

We construe the term organizational culture to indicate a set of behaviors that focus on the mindset that leaders, managers, and people in an organization are looking to apply systematically in the achievement of their stated missions. Those are behavior patterns that can be led and change managed.

Often mindset cues are found in organizational purpose or vision statements. For example, the Virgin Group's purpose is "changing business for good."[14] Richard Branson, the Virgin Group's leader, lists his organization's main behavioral expectations as "happy people doing their bit," being "well trained" and "motivated."[15] Motivation levels can be measured and the organization can take remedial actions to support people who lack motivation.

Ethical, or unethical, decision-making indicates people's mindset and an organization's culture. Digital-era technology has virtually limitless potential, including the ability to harm organizations and people if not appropriately used. Digital systems cannot make ethical decisions; these automate the decision-making process described by the programmer.

Senior Uber managers were excoriated in the press for permitting unethical risk-taking. The managers used a proprietary technology, "God View," to track users' ride data and to retaliate against journalists.[16] These ethical breaches were well known within Uber – in fact, they were encouraged and celebrated at company parties – and there was little consequence for Uber managers' poor choices.[17]

"Good management is about making choices, so a decision not to do something should be analyzed as closely as a decision to do something."[18] Organizations can choose to praise those managers who note a digital-era technology's possibility to undertake an unethical practice, say monitoring female journalists' personal lives, and who make a better choice. And, to reinforce good decisions, organizations should hold their people accountable for making bad ones. When managers have violated an organization's norms, these failures should be given no safe harbor.

We include in behavior change the idea of encouraging and reinforcing mindsets. Behaviors can be change-managed using the playbook to digital era change leadership presented here. Encouraging positive behaviors, including the norms that align people and process with technology, can establish a digital-era mindset that will help an organization achieve its mission.

BUILDING DIGITAL ERA BEHAVIOR CHANGE CAPACITY

When people view every new change initiative as a threat to their knowledge, power, status, or way of life, they can become change martyrs. People become cynical and distrustful. Much like the knife-thrower's assistant in the circus, they are constantly trying to dodge the latest change effort. When in a constant state of fear, an individual cannot summon the physical, emotional, and mental resources to deal with change.

Technology takers build their capacity to continually change. After all, we have become accustomed to automatic smartphone updates that keep us current, even if we have not invested in the latest model. The constancy of technologically induced behavior change is second nature to being part of the digital era.

The challenge with managing the adoption of digital-era technologies and related behaviors is not change resistance per se. Rather, it is factoring in the constancy of change *and* recognizing how digital-era change management is an iterative, if not infinite, game. Kotter's definition of organizational transformation as a process is expanded, in the digital era, to be a never-ending story of constant dynamic change.[19] For an organization to succeed in the digital era, the whole of the organization and its people must lead technology-taking "at all levels, in all locations."[20]

Leaders who choose technology taking as a strategy choose to embark upon an exciting journey. This trip has many twists and turns. We offer a playbook to guide the adventure, to avoid some of the pitfalls while reaping the benefits of increased business value.

This requires investing significant time and effort in building organizational capability and capacity to change, and both envisioning and creating a climate that seeks, rather than rejects, behavioral change. For those organizations prepared to rise to the challenge, we outline five plays for technology takers to influence or change behavior and succeed in the digital era. The plays support a virtuous cycle of change management driven behavior changes to match the requirements of the digital era. The cycle is continuous, and each time an organization moves through the circle, it builds capacity for the next trip. Unlike a vicious circle, where everything gets worse, in this virtuous circle, the organization continually improves its technology-taking capabilities. A virtuous cycle of change allows the technology taker to adapt to the digital era and adopt the technologies that deliver value to the whole organization.

The plays are not exhaustive of digital era-focused, behavior change management actions that an organization could take. Instead, the playbook is meant to represent a starting point for organizations that want to play as technology takers.

Figure 3: The Virtuous Cycle of Change.

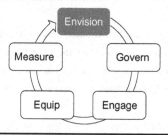

VIRTUOUS CYCLE OF CHANGE

Figure 3 represents a virtuous cycle of digital-era change management. A cycle is a set of events or actions that occur again and again in the same order. Navigating the circle by completing the five plays makes it easier to do so the next time. At first, organizations focus on a single technology implementation. A successful technology adoption with resulting behavior adaptation builds confidence and capacity for the next one. Soon, an organization will recognize that change is continuous and will develop change management skills to make each implementation more successful than the last. Rather than treating each change as separate and distinct, an organization will manage a portfolio of change (Figure 3).

GUIDE THE JOURNEY WITH FIVE PLAYS

Excitement is always strongest at the beginning of the technology-taking journey; and many times, organizations simply want to get started. This is the time to capture all of that excitement into a compelling business case and vision. Just as NASA astronauts could proudly point to the moon as their ultimate destination, technology takers need a beacon to draw them, both intellectually and emotionally.

As a first play, the organization should *Envision* itself as a digital-era organization based on a sustainable business case created at the highest levels of the organization. The purpose of this business case is to clearly define why the organization will choose technology taking. This business case cannot exclusively enumerate the benefits of technology taking. Many other initiatives compete for top management attention, all with a compelling return on investment projections. In order to keep focused, leaders must agree that delaying or refusing technology taking will cause irreparable harm. The risks of staying the same are far greater than the risks of getting on the technology-taker bandwagon.

To avoid staying the same, senior management must have a business case for and explicitly express ownership of technology taking as its strategy. Technology taking is not a bottom-up process but must be led from the top via delivery of a constantly updating business case that documents implementation of the technology taker strategy. The business case will set out how the organization will adopt digital-era technologies and will change manage its people's behavior in response to the technologies' demands and processes.

As with all good business cases, it must be envisioned as a future endeavor. It cannot include sunk costs of past processes. Already electronically enabled processes are of the past and should not have any bearing on future decisions for the business case.[21] The business case envisions changes to behavioral mindsets for people and processes that are driven by technology. Sustaining a virtuous cycle of change is not left to coincidence. To do the heavy lifting, including developing and constantly updating the technology taking business case and vision, a new *change management function (CMF)* is created.

The new, robust CMF will implement the technology taker strategy per the business case. The CMF will help senior management overcome change fatigue and provide both upwards and downwards leadership on the playbook's implementation. The CMFs job is to develop a package of updated change management interventions that respond to technology-taking innovation and value creating opportunities, including related behavior or mindset changes.

Once the beacon of the technology-taking vision is shining brightly, employees will begin to look that way, and many will begin to take steps in that direction. But not all steps will be appropriate. Looking at the organization's mission and values, leaders provide guardrails to prevent people from stepping into dangerous territory and provide a sense of security and safety during the journey. These guardrails take the form of policies and procedures.

Having envisioned its technology-taking, the organization will need help to be ready for the digital era. Described at Play 2, readiness implies assessing an organization's capability to change and setting out the structure the organization will use to *Govern* and effectively guide change. The early stages of any change are highly ambiguous and confusing. People wonder, "Is everything changing? Is anything staying the same?" Consequently, wise leaders underscore what is staying the same: organizational mission and values. They also need to provide tangible evidence of what is changing. By tackling governance issues early on, leaders signal that they are serious, as they are spending time and energy to address and properly institutionalize issues that will arise once technology taking is underway.

The organization's mission, its policies, and its procedures are often forgotten when technology adoption occurs as a project and not as a corporate-wide, iterative initiative. Organizations may attempt to govern technologies or change management according to separate policies, procedures, and structures of responsibility. Or the envisioned business cases' implementation is not governed for the digital era. Technology taking will not spread throughout an organization until its governance addresses continuous technology adoption and adaptation.

With the beacon shining and the guardrails in place, leaders invite everyone to participate in the journey. Some will jump in, others will hold back. And some will begin, only to fall away when they get tired, distracted, or scared. To complete the journey successfully, technology-taking leaders ensure that all participants catch the excitement and move along the path. Doing so requires visible, active, committed sponsors and advocates waving their flags so that everyone can see and join in.

At Play 3, the CMF will help leaders *Engage* with stakeholders about the future state envisioned and about specific business cases for technology adoption. For change to be successful, the leaders of an organization must *sponsor* the transformation desired. Sponsors mandate change by explaining why, what, and how people will be affected as technology ascends the Behavior Change Delta and overtakes organizational processes. Advocacy supports sponsors to inculcate change in an organization. Advocates use the very technologies for which they advocate to inspire others about the value of the digital era and to demonstrate the changes required to advance to the digital era.

Engagement helps people wrap their hearts and minds around the changes to come. Through direct interaction with sponsors and advocates, employees walk down the path of mentally and emotionally processing how their lives will change. For some people, this will be a long arduous journey. Successful technology takers recognize that helping people grasp the personal implications of the change builds trust and respect, and help speed adoption and adaption.

As people prance down the path of technology taking, leaders must ensure they are equipped to stay the course. Some may already have appropriate skills, but for others, this may be completely new territory. Training becomes critical. Even those who start the journey with a high level of expertise will encounter new terrain. So training must be continuous.

It is time for Play 4, *Equip*. Each technology has a methodology for its implementation, but the common aspect for digital-era technologies is the need for *training* on how to adopt them. It is essential to equip the people who are adopting and adapting to new technologies and ways of working. Technology-taking organizations recognize the importance of ongoing, continuous training at all levels. While managers may need a different approach than first-line employees, ignoring any group will impede technology-taking progress.

The CMF leads investments directed at the changes in the technology, process, and people aspects of the Behavior Change Delta. Specific investments in training will explain how to use constantly updating processes that are being common to all users of technologies. Investment in equipping people too will help change behavior as required by the technologies used.

As highly trained and equipped employees follow the flags waved by sponsors and advocates, they will continue to adopt and adapt to technology. Sometimes, they may appear to be moving sideways, and sometimes they may appear to be moving backward. The guardrails of policy keep them from straying, but how do leaders know if progress is being made? Progress is seen while passing the mileposts along the way. Leaders must define the mileposts and observe when they are reached.

Finally, *Measure*, described in Play 5, is a cardinal aspect of the virtuous cycle of digital-era change management. Readiness, engagement, and actions are measured, as are the overall achievement of technology taking and value created. Play 5 reminds organizations that an indispensable aspect of *measurement* is managers' own technology adoption and adaptation. Technology taking will be successful when an organization's leaders change the way they work and use the technologies adopted by the organization. Managers must show themselves apt at executing via information from digital-era technologies, instead of controlling the movements of people and the minutiae of processes.

Anything that is not measured or monitored does not receive adequate management attention and focus. All the great speeches in the world will not change organizational behavior. When leaders firmly understand the importance of technology taking to creating business value, they will not hesitate to measure progress and hold people accountable for changing their skills and mindsets. Without measurement, management communication is lip service.

When deciding to become technology takers, leaders and organizations must change-manage the intrinsic features of twenty-first-century technologies to transform people's behavior. Using the five playbook plays will facilitate implementation of the technology taker strategy. The playbook for technology-taking will also guide behavioral adaptations to technology and will lead to value creation as plays are used in an iterative fashion.

The cycle is continuous. Having completed one circumnavigation, leaders celebrate progress and learn from the experience. They then prepare for the next cycle by starting anew and revisiting the vision.

PLAY 1: ENVISION CONTINUOUS CHANGE MANAGEMENT

The Play: *To make technology taking a possibility, seek a leadership vision and support from a new change management function.*

A leader's most important attribute – the one that separates leaders from nonleaders – is "envisioning exciting possibilities and enlisting others in a shared view of the future."[22] Truly great leaders understand how to put facts into context, which gives them vision, meaning, and purpose. Leaders motivate others by inviting them to join in their purpose.[23]

The CEO of a large European retail bank developed a compelling vision by using clear metrics and gaining buy-in from his managers.[24] First, he set clear targets: doubling the economic profit of the bank, reducing its cost-to-income ratio to 49% (from 56), and increasing its annual revenue growth from the current, 1–2%, to 5–7% – all within four years.[25] The only way to achieve these goals was to empower individual managers to make rapid decisions using the information available to them.

Next, the CEO drafted a story to describe the bank's current position and the desire to reduce bureaucracy and increase entrepreneurship through decentralized information and decision-making.[26] His executive directors reviewed and refined the draft, taking ownership of the story.[27] Each of them created a chapter further illustrating the situation and potential changes in their department.[28] Every manager created a performance scorecard defining measurements of goal achievement. Halfway through the program, the bank was halfway toward meeting its targets for reducing its cost-to-income ratio and increasing its revenue and economic profit.

The retail bank example illustrates an element of vision-casting: defining success measures at the outset. The scorecard of the cost-to-income ratio and revenue growth was straightforward and understandable. And each manager had a personal scorecard measuring business results through the use of information and rapid decision-making.

The case helps visualize technology taking as a vision-driven process of constant behavior change. Except, for the technology taker, behavior change is also driven by the need for a leadership vision necessitated by digital-era technologies. Technologies that are now fundamentally changing entire industries. Leadership in the digital-era remains unchanged in its responsibility for setting out a clear, measurable, and relevant vision for success.

Change leadership is contingent on a vision of technology taking, as well as the capacity of the organization to carry out the vision successfully. With respect to the continuity of change, Play 1 addresses the need for change leadership within a virtuous cycle of change. After the initial envisioning is completed, the other plays of the playbook turn to governing change, engaging the organization to make technology taking work, and investing in and monitoring all aspects of change management.

ENVISIONING CHANGE

At the beginning of an implementation plan for the technology taker strategy, leaders must write down a vision for technology taking and provide the leadership to implement it. Envision as a play involves senior leadership arriving at a position of unequivocal ownership of the organization's technology taking. Leaders cannot become change leaders nor expect it of others unless a clear vision has been established. Vision ownership is not the same as change sponsorship, which relates more to authorizing specific behavior changes (see Play 3). Rather, the creation of change leadership is dependent on senior leadership owning the technology-taker change vision, like the bank CEO owned his business case for making his bank profitable.

In the past, many large projects would be done without project managers.[29] Organizations then slowly realized that project management was essential and many organizations that now deliver projects have certification requirements for project management, both at the individual and organizational levels.[30] Risk management too has followed a similar trajectory, in that risk specialists now are certified and involved in efforts to catalog and respond to risks.[31] To implement change in the digital era requires leaders who are knowledgeable about change management and who can rely on a playbook specifically designed for digital-era change management.

The Envision Play introduces common leadership ideas that need to be present in an organization for any strategy to be implementable. An organization's leadership envisions how it will meet the dislocations of the digital era. From this vision, a business case is developed to defend the selection of the Technology Taking Strategy. Although leaders set the vision, they cannot achieve it on their own. We observe many leaders stopping short on investing in change management after having made a decision to use technology as a driver of change. Setting out a vision and making a decision for change is not enough in and of itself; the journey ahead is long and ambitious.

Traditionally, external consultants or strategy teams would amend the business case to address a wider set of issues, including industry-level competitive forces.[32] Those were valid approaches when technologies enabled internally developed processes or human resources plans. Now, an internal CMF should ensure the primacy of technology when implementing the organization's modern vision. The CMF serves the role of navigator for the leaders. The CMF team includes people who understand the journey of change and have traversed it before. The CMF can see around the corners to anticipate what is ahead. In addition, members of the CMF team have a strong grounding in the current organizational structure, politics, and culture, so they know how to anticipate resistance and build support. The CMF unites and coordinates the previously disparate functions of change management teams, strategy teams, project management teams, governance committees, and change agents. CMFs build commitment for change among leaders, managers, and employees. Successful implementation of the technology taker strategy hinges on the CMF's ability to action an organization's vision for technology adoption and behavioral adaptation.

The World Food Program (WFP) is currently leveraging blockchain technology to track school lunches in Tunisia so that each child there is fed.[33] WFP previously used the blockchain to transfer vouchers based on cryptocurrencies to refugees in Syria, covering more than ten thousand people and generating significant savings.[34] Given the complexity of the technologies involved, the profundity of behavior changes necessitated, and the difficulty of implementing a vision to exploit modern technology, WFP developed a CMF to support its adoption of digital-era technologies like the blockchain and cryptocurrencies.[35] As a United Nations agency, WFP has to follow a myriad of rules and regulations that make innovation difficult. If WFP can create a CMF, any organization that wants to join the digital era can do so. As will be outlined further, a new CMF can solidify a constantly updating business case and help leaders convert their vision into digital-era reality.

Develop a Business Case

Successful technology takers define a business case that specifies how success will be measured for both the strategic vision and change management. The business case can emerge when executives decide to adopt new technologies because they sense that the digital age is passing them by. Their business needs have outstripped their technological capabilities, and others in their industry have leapfrogged ahead. This sense of urgency can compel companies to spend money and launch technology-taking initiatives.

But it is not enough to create and sustain behavioral changes required to adapt to these technologies.[36] An effective business case defines and quantifies the business benefits of technology adoption and adaption. The initial business case offers the opportunity for negotiation. If the benefits are not strong enough to justify the investments – technology, process, and people – then the plan can be modified. Managers must understand that they will be held accountable for realizing the projected business benefits.[37]

As technology moves to the top of the Behavior Change Delta, the usefulness of a business case will also increase. Digital transformations can fail because "organizations tend to start by picking solutions they like the look of, rather than fully evaluating the business case."[38] The playbook describes the use of a cyclically renovated business case that supports investment where it is needed. "Envisioning" writes down in the business case why and how to respond to digital-era dislocations. The playbook then provides the practical ways to proceed, including measuring change management (Play 5). Once the plays are completed, an organization's leaders should review the business case and adjust it for further technology taking.

The business case's "return on investment" calculation should include those costs and benefits of implementing change management. The business case further should describe generating technology-related innovation and value and avoiding doing nothing to respond to the digital era. Investing in digital-era technologies is not without cost and cannot be left to chance. Addressing the expected returns on investment from technology taking should make it easier to envision and implement change decisions.

Change management should be anticipated in the business case for technology taking. Whenever the business case is revised, its edition should be informed by the lessons learned from change management and the ideas generated through the use of the playbook. The CMF will be the business case guardian and change chief navigator, but not the captain or owner, of the digital era technology-taking vision and business case.

FROM A CHANGE MANAGEMENT OFFICE TO A CMF

Only the most visionary technology takers have started to create a CMF. The WFP has established an Innovation and change management director in its executive office. The director leads the CMF, here called the Division for Innovation and Change Management, which leads the WFP to use mobile and big data as a catalyst for ending world hunger by 2030.[39] The Division manages the change process of adoption of new technologies into the practices of humanitarian aid delivery.[40]

Enabling-era Change Management Office

A change management office (CMO) is the more typical way of managing the adoption of new technologies (although, fewer than half of organizations undergoing change projects have developed a CMO).[41] The CMO is an addition to a standard project management office.[42] The project or strategy team will prepare the technology system for the organization or develop a strategy that defines a way forward for the organization. Senior management endorses the strategy. Then the change management team, possibly via a CMO, will work on preparing the organization for the specified changes, harnessing senior management sponsorship. The change management team does so by following a predefined change management methodology.

There is a clear division of labor between the project and change teams. The definition of new ways of working is done by the project team. The change team, sometimes through a CMO, leads the organization's adaptation to these specifications.

Digital-era CMF

Adoption of digital-era technology and adaptation to that technology are parallel processes that need to be managed by one team. The project team cannot do its work properly with change management as a secondary step executed by a separate, and sometimes lower-level, team. Because change is constant in the digital era, change management must be a permanent function. The CMF is different from the transient CMO. A CMO can grow into a permanent CMF, which coordinates change, adaptation to digital-era technologies, and management of the organization's adoption of technology.

A CMF must be composed of elements of both a strategic project team and a change management team. This combination of roles and responsibilities permits the CMF to horizon-scan for digital-era opportunities, build the case for adoption of new technologies, and guide adaptation to these technologies. In describing the changes to come, the CMF will not be able to compare the current state of affairs to the desired future. The future is unknowable, as the technologies that define it constantly change their processes. And the present cannot be controlled by the organization itself. In the digital era, new ways of working are driven by externally defined processes.

The CMF must be led by a member of the executive team, perhaps a Chief Change Officer (CCO). This can augment the organization's ability to arrive at better recommendations of which digital-era technologies to adopt and to adapt to. The CCO also can reinforce the organization's decision to use technology taking as a strategy.

Outsourcing Change Management Efforts

In the past, electronically enabled times, a proposed strategy, product, or process would be envisioned and supported by, for example, a customizable ERP. Then, a consultant would bid for the work. The organization would accept the bid, but not the bid's last two pages covering change management. Those were the expendable part of the plan whose elimination lowered the price just enough. The consultant's time would be spent on high-level strategy and implementation work, but not on change management required for actual change adoption.

In today's behavior-changing digital era, most proposals start rather than an end by outlining the importance of change management as a factor of success. Usually, the change proposals follow the classic model outlined at the beginning of Chapter 4. However, the outsourcing of change management does nothing for an organization's internal capacity to deliver change management continuously over time.

Change management is following the trajectory of risk management, which too began as an outsourced activity and over the past decade became an internalized function. The US Sarbanes-Oxley Act required public companies to incorporate effective systems of risk oversight. Following the 2002 enactment of Sarbanes-Oxley, risk managing organizations[43] and certification schemes[44] sprang up, ready to support companies in their compliance. The International Organization for Standardization developed ISO 31000 in 2009 to provide a uniform set of principles and guidelines on risk management.[45]

A company's risk management activities could be conducted in-house by employees of the company or outsourced to consultants. At first, given most companies' internal lack of expertise in risk management, risk review was an outsourced activity.[46] Internalizing risk management is associated with decreased cost, improved ability of internal people to undertake risk management activities, and more frequent conduct of risk management activities.[47] Organizations could enjoy these same advantages by developing in-house change management capabilities.

Benefits of an Internal CMF

Unfortunately, some organizations have attempted to outsource the entire CMF function to consultants. Of course, a CMF can utilize the expertise and skills of consultants. But when outsiders take responsibility for both technology adoption and adaptation, the organization may fail to fulfill important responsibilities, considering that success or failure now rests with the contracted experts.

For example, one international manufacturing firm hired a team of consultants to implement an ERP system in all divisions around the globe.[48] The project ground to a halt after $20 million had been spent. No one could explain the vision beyond the need for up-to-date software. The consultants were doing their job — attempting to implement an ERP and implement business process changes in support. Without an internal CMF recognizing that this was not merely a technology project but that it was meant to change the behavior of every plant, the project got off on the wrong foot. Once a CMF was staffed with the appropriate levels of responsibility, authority, and executive access, it spent 12 months meeting with the plant managers, vice presidents, and the president to develop a business vision of a global factory, where components begun in one country could be shipped to another country for final assembly. The most important measure of success became time-to-market for new products.

"The sheer volume of technologies, processes, and decisions required to build and maintain digital applications and operations means companies can't afford to work in the same old ways."[49] The CMF has a role to play in implementing decisions of strategic relevance, including any decisions related to digital-era technologies. The alternative is business as usual, risking executive decision-making without due regard for technology's benefits and drawbacks. That approach would lead to flawed decisions.

Having a CMF is a play for technology takers ambitious for success in the digital era. As mentioned, the CMF must be positioned close to senior management or with its own corporate change officer to be effective. A CMF will help the organization develop internal capacity to capture digital-era innovations that are unlikely to be forthcoming from line managers. The CMF does not require from managers extra work they do not have the skills necessary to perform. Instead, the CMF assumes responsibility for bringing to light new opportunities, technology taking, and managing behavior changes to respond to this technology.

CMF RESPONSIBILITIES FOR IMPLEMENTING TECHNOLOGY-TAKING

The CMF ensures an organization always uses solidly researched ideas, impact analysis, and metrics-based feedback loops. The CMF persists even in the face of change resistance or digital-era deficient projects from powerful line managers with the right connections. The CMF requires a clear function description that sets out its two core responsibilities: to manage the taking of digital-era technologies and to guide the organization's adaptation to digital-era technologies.

Manage the Adoption of Digital-era Technologies

The CMF's most challenging task is helping their organization obtain and use digital-era technologies. This is not a task that traditionally is part of change management. Nevertheless, the purpose and promise of digital-era technologies are to supplant outdated processes. Efficiency and effectiveness cannot be improved if organizations try to adopt the new and maintain the old simultaneously.

The CMF builds a digital-era change plan that includes technologies that will respond to the challenges of the digital era. With a CMF, the Chief Information Officer or managers will not be permitted to shop around for their own, unique, and functionally optimal, but organizationally suboptimal, solutions. Instead, the CMF can suggest to organizational leaders good digital-era technologies that match their needs. By selecting the digital-era technologies the CMF advises, the organization can generate savings from not having to invest in customization or process automation. The CMF will help the organization to stay current and manage available, transformative tech options.

In 2017, the UN Office for Project Services (UNOPS) organized a CMF, a joint working group that guides the digital-era technology adoption and adaptation of WFP, UN Development Programme (UNDP), the UN Children's Fund (UNICEF), UN Women, the UN High Commissioner for Refugees (UNHCR), and the UN Development Group (UNDG).[50] The work of this group has resulted in research and development of the WFP's blockchain solutions in addition to the WFP's unmanned aerial vehicle (UAV) project, which it intends to use for topographical data collection.[51] These technologies replace traditional approaches to providing school lunches, disbursing funds to refugees, and analyzing natural disaster data – many of which were digitally enabled. Innovating with digital-era technologies allows the WFP to gain from behavior changing technologies and deliver aid more efficiently by eliminating millions of dollars in bank transfer fees, in the case of using blockchain.[52] Adapting WFP processes to the digital era offers donors better "value for money."[53]

WFP shows that a CMF is a strategic tool for keeping an organization relevant in the digital era. The UNOPS working group helped WFP identify modern technologies to supplant its current systems. The CMF then demonstrated its features by assisting imagination of creative ways of using these technologies within the scope of the WFP mission. It generated for WFP savings and improved services through the implementation of the behavior changing technologies and attracted donor investment and sponsorship for these efforts.

The WFP example also stands for the proposition of the increasing strategic importance of technology. As IT is prevented from developing (technology-tailoring) in-house solutions to needs, the tension between the IT department and line managers is likely to increase. An organization's leadership needs to own the decisions now being affected by technology taking. The CMF can monitor managers' compliance with these decisions and ensure managers do not have the power to stop changes or to refuse to use the technologies the organization has chosen.

Coordinate Adaptation to Digital-era Technologies

The constancy of technology change makes it essential that the CMF has the capacity to link digital era-induced behavior changes to business goals. The CMF can support leaders' decision-making by ensuring that decisions are based on the synthesis and application of metrics to information obtained through data streams. The conclusions drawn from analyzing digital-era technologies' data support the adoption of technology-taking behaviors.

The information in the data stream is what matters. Digital registries hosted on a worldwide network of computers are being tested in many settings, for the potential to reduce bank wire-transfer costs.[54] WFP is using a digital-currency network, Ethereum, which allows issuing tokens that can be tailored to needs, including money transfers, register sales, or even tally votes.[55] Using the data from the tokens, WFP can simultaneously reduce bank fees and monitor food assistance payments to hundreds of thousands of refugees. Other UN agencies, such as the United Nations Development Programme, are assessing using the blockchain to improve elections security and to eliminate the potential for ballot tampering. A joint CMF is guiding several UN agencies through these potential options, how to understand them, and how to use them.[56]

A CMF ensures a strategic, rather than operational approach, in support of technology-taking, one that balances the organization's technology skills and understanding with the demands of the new economy. The CMF helps leaders manage staff's reactions to behavior-changing technologies. Further, the establishment of a CMF establishes a career path for technology takers, contributing to the long-term availability of change management skills within an organization.

CHALLENGES OF ESTABLISHING A CMF

While the reasons for a new CMF may be clear to those who embrace technology-taking, it is challenging to set one up. For the CMF to work, it needs to be empowered to do more than create a Kotterish "sense of urgency."[57] It must be able to move beyond garnering executive buy-in as the primary success factor of change management.[58]

The historical success rate of change management projects and transformations is, frankly, low.[59] Change frustration is why organizational leaders take on support for change initiatives. Formal leaders cannot do all the heavy lifting, though, and need help building a sense that real change is taking place. Without support from a CMF, technological adaptation is seen as an additional responsibility for the average worker or manager to assume along with her normal, daily tasks. But any CMF's credibility is challenged from the onset by change resistance, frustration, and exhaustion.

Instead of developing a CMF, an organization may try to put CMF tasks in its managers' job descriptions. Yet the infrastructure to carry out needed change management work seldom is sufficient. Being released with short-term back-up or pulling double duty as a line and change manager is a short-term solution; but, change initiatives can easily interrupt the flow of operations. With just-in-time production methods and administrative lean techniques to respond to, leading change management efforts is a thankless task. Managers also do not have the skill set required to be change managers. Like for risk management, change management is now a concrete discipline with certifications and coursework. Few have obtained certification as change managers. Those who have may be hired into human resources or communications departments at lower levels of seniority than what is actually required to effect organization-wide transformation.

ENABLE CHANGE WITH A CMF

The digital era's constant change cannot be addressed with the typical approach of having a project team develop a new business strategy for the organization and then having the change management team prepare the organization for those changes. For the technology-taking organization, a business case for the digital era must be executed by an internal CMF.

The challenge of setting up a CMF can be overcome by recognizing the function's responsibilities to support technology adoption and adaptation. With the support of a CMF, the Envision Play is a conscious investment in helping leaders understand the digital era so they can direct their organizations to become technology takers. The CMF engages with all the elements of the virtuous cycle of change: governance, engagement, investment, and measurement. A CMF serves as the switch through which leaders can provide change direction and management simultaneously, rather than as separate steps executed by project and change teams.

A CMF represents a potential upheaval in the way strategy and technology are usually sponsored, developed, and governed. The CMF has a role as both the secretariat of digital-era governance and the senior-level distiller of what to sponsor through engagement. Also within the CMF scope of responsibility is coordinating training on digital-era skills and measuring the capacity of managers to use technology. These elements of the virtuous cycle of change management are our next plays.

PLAY 2: GOVERN TECHNOLOGIES AND CHANGE

> **The Play:** *Establish governance structures to guide adoption of and adaptation to digital-era technologies.*

In March 2018, the US Federal Trade Commission, the US Senate Judiciary Committee, the US House Judiciary Committee, and the attorneys general of 37 American states initiated investigations of Facebook over its policies – or lack thereof – preserving the privacy of users' browsing and other personal data.[60] Facebook has admitted providing its users' data to a private company, which attempted to use these data to manipulate the user beliefs and votes in a US presidential election.[61] The Facebook debacle, currently threatening the company's long-term credibility, was, at its core, a failure of readiness for technology taking by its users, as reflected in the poor governance of those changes by Facebook.

WHY GOVERNANCE MATTERS

New technologies usher in the new freedom of action. Employees can easily do what was previously expensive, time-consuming, difficult, or impossible. Technology takers like to experiment – if it can be done, someone will try. As long as these experiments are consistent with the organization's mission and values, they should be encouraged. When they are inconsistent with organizational mission and values, they should be proscribed. Properly decided on, documented, and applied, policies make values explicit – people understand what they can and cannot do, which supports an organizational culture of continuous change.

Leaders need to stay ahead of the game to avoid nasty surprises. In some cases, people will not attempt certain actions because they are prohibited by existing policies. For example, while workers could use technological tools to work from home, most will not risk running afoul of workplace policies that specify their office as their sole worksite for the protection of trade secrets. However, if freed to telework or to work remotely, people who are accustomed to securing information with a lock and key may not even consider the consequences should their smartphone fall into the wrong hands. Policies written to address physical assets are inadequate for virtual assets.

The risks of technology taking necessitate the second play of change management: establishing readiness for change through organizational governance.[62] Defined for the purposes of technology-taking, governance is a unified system of decision-making on policies and procedures that guide the technology, processes, and people of the Behavior Change Delta.[63] Governance through a hierarchy of policies and procedures will control the risks of adopting and adapting the digital-era technologies used by the organization. Governance involves the creation of a whole-organization governance structure, establishment of managerial responsibility for policies and procedures, and compliance monitoring.

Discussion of a new CMF in Play 1 showed that a business case always underlies the adaptation and adoption of digital era technologies. Policies and implementing procedures will set out how the business case will be achieved consistent with the organization's mission. Governance supports organizational change readiness because it aligns capacity building and communication with the organization's guiding mission.

CREATE ONE ORGANIZATIONAL GOVERNANCE STRUCTURE

Technology takers live in the tension between centralization and decentralization. With more and more information available at the fingertips of first-line managers, most organizations are moving toward decentralized decision-making.[64] The role of governance is to ensure that these individual decisions are putting into practice the organization's mission and values – commonly held beliefs supported by centrally driven policies. Governance, therefore, is a centralized function with authority over all departments and functions.

Governance helps organizations respond to the digital era's challenges, including the incredible pressure to develop separate governance structures for newfangled technological systems. To those who are not technology takers, an ERP system appears merely an administrative tool, having no bearing on the mission or "true work" of an organization. Rather than trouble managers from other business lines, organizations may task the IT department to decide matters of governance related to that ERP system or any other particular technology. Or a tech committee may be convened to govern the approval of IT system changes.

While potentially easier initially, separate governance serves to limit digital-era technology adoption into and adaptation by all aspects of the organization. This limitation constrains the effectiveness and the behavior changing potential of digital-era systems. Separate governance structures also create a disjointed and ultimately ineffective decision structure for the maintenance of an organization's policy and procedural framework, one that can threaten compliance and control throughout the organization. Further, different rules for different departments permit managers to make decisions that are ideal for certain departments but suboptimal for the rest of the organization.

The systematic generation of policies and procedures, along with their permanent storage in an accessible and reusable fashion and the ability of skilled people to cross-reference and make connections between them, are fundamental capacities for organizations in the digital era.[65] Therefore, a modern organization's governance structure should contain four elements: a guiding mission, a single governance committee, realistic policies and procedures, and a unitary policies and procedures manual.

Refer to the Organization's Mission

An organization's mission is its lodestar, that which guides its governance decisions and serves as the ultimate check on the demands of technology-taking. Several United Nations organizations have recently adopted digital-era ERPs to serve as information management systems.[66] These UN organizations established and published clear expectations to strengthen the governance of their organizations generally and their ERP systems specifically. Any ERP system had to assist the organizations to achieve their missions in human rights, health, or development in a more efficient and responsive way. The ERPs did not supplant the organizations' missions; they were merely tools for better achievement of missions.

Good governance supports the organization's business by aligning policies and procedures for achieving the organization's mission. Google's employees protested their company's contract with the US military to use artificial intelligence to improve drone capability. Although Google's first motto was "do no evil," the company had no developed policies that specified the rules of its engagement with the military. Following a revolt from 4,000 employees, Google declined to renew its contract with the Pentagon and committed to developing a new policy to preclude the use of A.I. in weaponry.[67]

Yet the rapid pace of digital-era transformation requires that organizations' governance is stretched to address situations never before contemplated. In 2017, Facebook had changed its mission statement to "bring the world closer together." However, the company's data breaches occurred during its previous mission, which intended to make the world "more open and connected."[68] If Facebook interpreted its prior mission to make open and publicly accessible the lives of its two billion users, this mission seems to have been achieved. If, however, Facebook had intended to keep users' personal data private, then the company failed to meet its mission as it was swept along in the very technological tide it generated.

Convene a Governance Committee

The next aspect of an organization's governance structure is its governance committee, which assumes ultimate responsibility for all organizational policies and procedures. Facebook is controlled by its founder, CEO, chairman and 60% voting authority holder, Mark Zuckerberg,[69] who sits "in California in an office making policy decisions" for billions of people around the world.[70] Yet, a governance committee of one may not be the most effective – or ethical – structure to manage the privacy decisions for billions.[71] A potential future approach for Facebook could involve giving users, whose data are collected by Facebook, more direct power to decide what the company can do with that data and how it will be accountable for breaches.[72]

A governance committee mediates disputes over the meaning of policies and their integration with processes. It manages the policy approval process and approves all policies. The committee also is the oversight body for procedural documents to ensure consistency with established policy. For example, rather than several, potentially inconsistent policies for telework, the governance committee ensures a single, unitary approach.

Significantly, the governance committee divests from individual managers or departments authority to establish policy or unique procedures.

In adopting its ERP, UNDP followed exactly this approach for governance. The organization's governance structure had to enable adequate coordination between stakeholders and allow for UNDP to make effective decisions within the organization's own hierarchical accountability structure. UNDP established a "Trilateral Advisory Panel," charged with recommending decisions related to the implementation of the ERP technology. However, this body's recommendations were then confirmed by the Senior Management Team, the primary internal governance body of UNDP.[73]

Governance committee members must be cognizant of and understand any technology modernization efforts that the organization is undertaking. Provided she has authority for technology adoption and adaptation decisions, the CCO should serve on the policy committee. The CCO would assist the policy committee to make certain that procedural changes, whether proposed or required by cloud-based software revisions, do not conflict with the organization's policies or mission. When an automatic revision to a business process is out-of-step with a policy, the policy committee must consider changing that policy. The CMF can ensure alignment with the virtuous cycle of change management.

Establish Real-world, Digital-era Policies and Procedures

Just as modern technology is disrupting businesses and economies, it is also disrupting governance.[74] Old policies, the rules of the organization that establishes authority, responsibility, and accountability, and existing procedures, those processes used to implement policy, are no longer sufficient to inform and describe the organization's response to the digital era. For change management purposes, policies and their implementing procedures must reflect the real circumstances in which an organization finds itself. For collaboration on the technology taker strategy, governance must help guide people's reactions to this reality. An organization must ensure governance of all aspects of the Behavior Change Delta: technology, people, and processes.

Facebook's policies governing data privacy have been described as "lax."[75] Company policy did not prohibit data harvesting from users, thereby permitting users' contacts, external entities, and even Facebook itself to collect reams of information about each user's biographical profile, geographical location, friends, likes, personality, beliefs, and desires.[76] Facebook policies required the individual user to shoulder responsibility for protecting her own data – even if the user was unaware of which data Facebook had collected or had permitted to be transmitted to others.[77]

For policies, the organization's governance committee should take charge of identifying and systematizing all organizational policies and drafting new policies to cover emerging or other areas ungoverned by any current policy. To be effective in its duties, the governance committee must take a risk management approach to policy development. That is, the committee should appraise the likelihood and severity of a possible threat to the organization.[78] Then, a policy or policies should be written to constrain this risk, to assign authority for addressing it, and to set out responsibilities related to it. Laws, industry regulations, and ethics too should be considered.[79]

Where a likely and severe risk is foreseeable, it is a governance failure to leave a policy hole through which the organization could be adversely affected. When the *Guardian* newspaper informed Facebook of unauthorized data harvesting in 2015,[80] Facebook then developed no new policies to control for future occurrences of this risk.[81] Only in the wake of a massive data breach that enveloped the company in lawsuit, scandal, and possible government regulatory action, did Facebook's CEO, Mark Zuckerberg, announce policy changes to protect user data.[82] However, these new policies may be too little too late.

Similar to Facebook, Uber too failed to protect users' data. Senior Uber managers used a proprietary technology, "God View," to track users' ride data. To retaliate against journalists who published critical reviews of the company, Uber followed the journalists and used ride information to bully them and their families.[83] For amusement, Uber managers recorded users' trips that suggested they were having "one night stands."[84] When outed in the press for these abuses, Uber pointed to its policies that permit collection of user data for "business reasons," as defined by the company itself. Unconstrained by policy, Uber had exploited the full abilities of its technology.[85]

The Facebook and Uber debacles demonstrate why, in the digital era, an organization's governance must also include policies governing data. Most countries have laws that require minimum data protections for organizations in sensitive industries, such as the legal, health and education sectors or doing business with certain customers.[86] Yet data stream-producing technology has moved faster than legislators, and most organizations must determine for themselves how to avoid the risk of reputational damage resulting from illegal or unethical data use.[87] Organizations must devise policies on data access, sharing, use, retention, security, and disposal. Public outcry over misuse of personal data has pressed organizations to go further than what the law requires and to consider the purposes of data collection, transparency of use, and owner's authorization and consent for their data to be shared.[88]

Implementation of modern technology also gives organizations an opportunity to streamline processes so that they align with best practices.[89] An organization's procedures should describe both processes particular to a certain technology, such as those used by an ERP system for hiring employees or transferring funds to departments, and processes that are wholly separate from technology. Standard operating procedures that describe digital-era technologies must be regularly reviewed and revised to match the constant system updates of cloud-based systems. Outdated organizational processes (or those described in 20-year-old paper memos stained with coffee rings) can be replaced with digital-era processes that conduct tasks more quickly and efficiently.

Readiness through governance also prepares the organization for one of the strategic challenges of technology taking, the increase in interconnectivity with other users of the same technologies. In the digital era, an organization's policies and procedures can affect external stakeholders. Twitter, for example, changed its policies to prohibit users from tweeting threats based on another user's race, ethnicity, national origin, sexual orientation, gender, gender identity, religious affiliation, or age.[90] But Twitter trolls exploited holes in the company's policies to abuse disabled people. Belatedly, and after lobbying from advocates, Twitter added disability to this list.[91]

Develop a Unitary Policies and Procedures Manual

A final element to a modern organization's governance structure is a policies and procedures manual. To control and regulate their approach to technology taking, organizations must corral their policies and procedures in a lex codex, a manual that is the centralized and authoritative source of the rules and processes the organization will use to govern itself. The policy manual establishes a hierarchy of interrelated policies, implementing procedures, and forms or other documented discrete business processes used to support the achievement of the organization's mission through technology implementation. When there is a conflict between documents, the instructions in the higher document will control. There is no stand-alone policy document because of the integrated and interconnected way that various technologies and policies can affect internal and external stakeholders.

The manual helps manifest the governance concept that no process, including any related to an IT system, exists without a policy basis. Although certain digital-era systems, such as SaaS ERPs, arrive with their own process documentation, a link must be found to join these procedures to an organization's policies, its business, as well as the technology-taking business case. The organization's policies will constrain its digital-era IT systems, for these systems are merely a set of procedures used to implement policies. The manual thus drives digital-era processes into the organization's governance structure.

Governance helps manage choices, which technology taking provides in abundance but that, through interconnectivity, affect all in an organization. The manual shows the organization's selection of particular procedures to implement the technology-taker business case. As part of an organization's technology and change management governance, a policies and procedures manual indicates which decisions are best for executing the technology-taker business case.

FOSTER MANAGERIAL RESPONSIBILITY FOR POLICIES AND PROCEDURES

Harmonized policies, procedures, and governance structures can calm turbulent waters among managers, who may be particularly vexed or threatened by their organization's adoption of digital-era technologies. Previous to an organization's adoption of, say, an SaaS ERP or Customer Relationship Management (CRM) system, middle managers' power came from their control over parts of a process, systems, or, in the CRM case, the ubiquitous rolodex of sales leads. As a result, business owners of specific processes would issue conflicting directives, leading to pitched conflicts among managers and employees who each owned a piece of the same process. Governance of digital-era systems ensures departments or managers cannot make decisions that favor themselves but that are actually imperfect for the organization. Governance supported by the transparency of data streams prevents it.

Make Managers Policy and Business Owners

Managers' resistance to new policies and procedures is overcome by increasing managers' governance responsibilities and ensuring work across departmental silos. Managers are made responsible for the accuracy of policies and procedures related to their functional areas, as managers are both policy owners and procedural business owners. Policy owners ensure all existing policies are incorporated into the policy manual and that all policies under their purview remain current and correct. They also may propose development or revision to policy variances to match the technology they are taking for the organization.

As business owners, managers are responsible for digital-era and other processes to implement their policies. Tasked with understanding or devising all the processes necessary to implement their policies, managers are made to use the technologies the organization has adopted. As they use the technology given to them, managers maintain and revise the procedures affecting their lines of business. If a cloud-based system, say a SaaS ERP or CRM, is used to perform a particular task and that system undergoes an update, the manager is answerable for the accuracy of the procedural documentation describing how to do a particular task. The description of the procedures must change along with the procedures themselves.

Empower Business Process Experts

SAP, the German software company, may have been the first to propose the development of the role of Business Process Experts (BPEs) to test proposed software enhancements.[92] BPEs are lower-level employees who assist managers in the conduct of their duties as policy and procedure owners. The BPE must remain aware of and test procedural simplifications, especially those affected by automatic updates of a cloud-based operating software. The BPEs have primary editorial duties to ensure that procedural documentation describes the most efficient process possible, whether in or out of the ERP, for implementing a policy. They also draft proposals for innovations or new ideas to emerge from horizon-scanning technologies at the procedures level. Those ideas are then fed into the business case envision play via the CMF.

The BPE should be convened as a group outside the authority of any individual department. Establishment and empowerment of the BPEs move final accountability for procedures from individual departments to its governance committee. This arrangement permits BPEs to work together to harmonize technology changes across departmental lines, thus reducing instances of conflicting procedures and internecine battles among departments.

As a fundamental principle of digital-era change management, organizations' efforts to rid themselves of old processes that could be supplanted by new technology must also resist internal efforts to maintain and use both old and new processes simultaneously. The Technology Adoption–Adaptation Matrix in Chapter 2 demonstrated that the digital era is too different from the enabling era to attempt to use the processes of the past to achieve the digital future. The BPEs unite understanding of an organization's business processes with digital-era technology applications. For the modern era, BPEs must assume new duties to be aware of technology-driven procedural changes and to test proposed procedural changes to ensure these are correct.

MONITOR COMPLIANCE WITH POLICIES AND PROCEDURES

Digital-era technology requires constant procedural change and perpetual policy vigilance.[93] Introduction of digital-era systems is an opportunity for an organization to build capacity for change, as it requires people to modify their behavior with respect to how they do their jobs. Improvements to policies and procedures and knowledge transfer activities are central components of both change management and governance inculcation throughout the organization. Capacity must be built for the organization to use data from modern processes for management of the organization's mission.

Organizational change to rely on policies and to use new procedures will be achieved through monitoring compliance with policies and procedures. Compliance is achieved through training, especially on policies and procedures; communication, including centralization of announcements of policies and procedures; and audits and analysis of data on work conducted according to the policies and procedures.

Although appropriate policies were in place, Wells Fargo employees nevertheless manufactured fake accounts in the names of existing customers, causing many to pay fees.[94] Wells Fargo lacked audits and other appropriate controls to monitor employees' compliance with the company's policies and procedures. The one policy employees paid attention to was incentive pay for number of accounts opened.[95] Managers pressured their employees to meet such performance goals with no review of how these goals were met. Employees admitted, "when you are under pressure to meet a tough goal, you do what you need to do, even if it's fraudulent."[96] In attempting to correct the problem, the company fired 5,300 employees and lost its CEO, John Stumpf.[97] The Wells Fargo example points to the need for more effective change management, to reduce stress and anxiety in times of change, and more effective governance controls, so that policies are not ignored. Those principles hold even without technology in the picture; but technologies of the digital era are dislocating for a reason: they are more powerful than what most organizations have been accustomed too. In the digital era, consequences for poor governance or compliance will reverberate forcefully.

Train On Policies and Procedures

Many do not understand the distinction between policy and procedures, nor why they matter for a strategic approach to digital-era dislocations. An organization may need to educate its personnel on its governance approach, manual of policies and procedures, and monitoring and implementation by the CMF, business owners, and BPEs. On-line, in-person, and video-based training sessions can be used to build capacity to change as required by policies and procedures. These sessions should distinguish between policy and process and explain how and by whom policies and procedures could be changed. Standard operating procedures serve as training materials, explaining discrete business process steps used to achieve policy mandates. Training sessions can provide opportunities for groups not traditionally involved in governance to work with BPEs and business owners to develop procedures.

Communicate

Digital-era processes are continuously revised, with major system updates occurring regularly.[98] Policy owners and BPEs have an ongoing task to describe changed processes and to ensure no conflict with policies. Communication of changed processes and policies is fundamental to helping people comprehend and follow an organization's governance mechanisms. This is no small matter; for, now that it is at the top of the Behavior Change Delta, technology will still be an unexpected source of change for most people and managers.

For significant policy or procedural changes, an organization's leadership should present a unified front. Senior management members from various parts of the organization should jointly announce departures from previous policies or adoption of new procedures whose effect will reverberate across the organization. When Facebook changed its policies to protect user data, the company's CEO announced these changes in a blog post, which was reposted and amplified four minutes later by Facebook's chief operating officer.[99]

The organization's policy committee centralizes announcements of policy and procedural changes, ensuring that all people receive timely and intelligible notices of what has changed and how. All procedures and policies should be announced via a single format or per a dedicated channel. Formalized announcements of procedures serve to establish these documents as definitive and management-sanctioned statements of a policy implementing instructions.

Digital-era technology adoption has provided organizations means to communicate in more modern, innovative, and targeted ways. The digital-era itself is allowing organizations to develop training courses, pop-up reminders, video messages, and reports on everything from messaging systems to official blogs. An organization's change and governance leaders can leverage these capabilities to reach all people with a need to know about new policies and revised implementing procedures. Change management does not happen in a vacuum; people need to know what the parameters are or they will become risk-averse and do less than what is required for technology taking to take root.

Analyze Data and Audit Behavior

A final aspect of compliance is an analysis of data on work conducted according to policies and procedures. The policy committee should collect and review data about how the organization's employees are working via the technologies the organization has adopted. Digital-era technologies' data streams can be used for this effort.

Information collected from these data streams show which workers are using which processes and to what effect. A UN organization adopted a cloud-based ERP, requiring old procedures to be abolished and new ERP-based ones established. Mining the ERP's data stream provided answers to whether people had changed how they were working and whether new procedures were being used. The results of this analysis informed training programs that were targeted to specific people that were shown to have the most difficulty with change.

An audit is another way of reviewing workers' adoption and adaptation to digital-era systems. Auditors review compliance with an organization's policies and procedures and provide independent assurance that an organization's risk management, governance, and internal control processes are operating effectively.[100] The purpose of these reviews is to help organizations succeed and to ensure the organization's processes are appropriate and are appropriately used.[101]

Government regulators and donors requires audits of organizations big and small, public and private, national and transnational. All stakeholders want to know whether an organization will succeed in the digital era and whether its policies and procedures truly govern behavior. Even tiny, nonprofit organizations, such as African Mothers Health Initiative,[102] a US$100,000 per year charity dedicated to reducing deaths of the world's most vulnerable babies and women, must report annually on their whistle-blower policy, their document retention practices, and their conflict of interest controls.[103] A successful audit of an organization indicates that the organization's governance controls meet its mission and that it has successfully navigated the digital era.

READY THE ORGANIZATION FOR CHANGE
IN THE DIGITAL ERA

Revitalized governance for the digital era means identifying what will change due to technology and relating these processes to the policies that achieve the organization's mission. Without comprehensive governance covering technology use and data analysis, organizations will be unable to steer the technology forces that affect them.

Managers will try to customize (i.e., tailor) systems to their own needs instead of adapting to digital-era processes. Governance for the digital era contemplates new roles and authority to implement, explain, or lead change. If properly created, a hierarchy of policies and procedures also will ensure transparent governance of the technology by the organization's policies and will limit instances of the digital-era tail wagging the technology-taking dog. Instead, the organization, through good governance, will become and stay ready to negotiate the virtuous cycle of change. Innovations can be codified in the policies and procedures manual to mitigate the risk that rogue technology tailors will use processes that are inappropriate to the organization's technology-taking needs.

Finally, successful transformation and modernization is contingent on an organization's ability to convince its people to use the organizational policies and procedures – instead of allowing managers to set up their own fiefdoms, each governed by a different set of rules and a different way of doing things – especially when it comes to whether technology-taking is a viable strategy. Building capacity for change and communicating about policies and procedures is integral to establishing compliance, thereby strengthening overall organizational governance, and to achieving desired changes in behavior, thus improving technology adoption and adaptation.[104]

The ever-evolving nature of digital-era technology forces organizations to be takers of processes. These processes are means to an end; and that policy end, whether written up or supported by a formal business case, is the organization's raison d'etre. And that remains absolutely under the organization's control. A strong governance play will ensure that, consistent with its mission, an organization will be ready to confront the challenges of the digital era. Once completed, it is time for digital era change leaders to engage to sponsor and advocate for change, which is the next play.

PLAY 3: ENGAGE TO SPONSOR AND ADVOCATE FOR CHANGE

The Play: *Sponsor, mandate, and advocate for change at all levels of the organization.*

In adopting their ERPs, the senior management of two UN agencies demonstrated how to sponsor the technology-driven changes that would modernize their organizations. The agency heads wrote and presented to all people a rationale[105] or mandate[106] for organizational change. These documents outlined why the organizations were implementing their ERPs and detailed what new ways of working and new responsibilities would be required of people. These mandates required the senior management team to be in full agreement about the goals of technology-taking. The joint presentations were a powerful visual message to people that the entire executive team supported the technology adoption and the changes that it would require.

The agencies also began training and empowering a new type of change agent within every department and country office. Beyond serving as conveyors of information, these change agents become champions and advocates for the new ERP systems. These advocates catalyzed change within their own spheres of influence and, ultimately, became the first-level support network before and after ERP go-live.[107]

Digital-era change cannot rely solely on leadership at the top of an organization. Sponsorship has to engage and cultivate change leaders across all levels and nurture new ideas about how to transform work behaviors and processes.[108] No matter where they sit in their organizations, technology takers must be ready to assume a role as digital-era change advocates. Play 3, "engage to sponsor and advocate for change" is the combined effect of sponsorship, a clearly mandated change direction, and advocacy for change at all levels of the organization.

MORE THAN SPONSORSHIP REQUIRED FOR DIGITAL-ERA CHANGE

Engaging, effective, and active involvement by senior leadership is traditionally believed to be the greatest contributor to success in change management.[109] "Sponsoring" refers to leaders welcoming change with open arms and explicitly communicating why change is needed. Effective sponsors legitimize the implementation of a change through influential communication and meaningful consequences.[110] When a sponsor commits to a change, she does so publicly, providing a galvanizing vision for change, and privately, supporting from behind-the-scenes.[111] The play to sponsor digital era change includes the clear communication of a mandate for change, which in turn has a clear link to the envisioned transformation of the organization. The mandate clarifies not just why and what is changing, but also communicates a clear direction, presumably toward the northeast of the Adoption–Adaptation Strategy Matrix to technology taking.

Despite the general acknowledgment that sponsorship is essential for change, most companies have a terrible track record in implementing organizational change.[112] And most organizations have dismal success with successful launches of digital-era technology projects.[113] In the digital era, something more than sponsorship is required for technology adaptation and adoption.

Advocates are the innovation for organizational engagement with the digital era. Usually, the cast of change characters is described in terms of sponsors, agents, and champions.[114] Not everyone can be a sponsor; but, any technology taker, no matter where she sits in the organization or whether she is an agent or champion, can be an advocate. Advocates are the super-agents or the ultrachampions.[115] They are personally committed to technology-taking. They see the benefits for themselves and the organization, and they are willing to take a personal risk. Often advocates remind sponsors why continuous change is required. Finally, advocates borrow techniques from lobbying and social marketing and apply these to change management.

Building a cadre of advocates follows a model of horizontal or pseudo-social networks of distributed change leadership.[116] This approach allows for change leadership by a broader group of people at different levels and locations. Creation of a community of advocates builds on the new forms of engagement popularized by some of the dominant social media technologies of the digital era. The existence of a written mandate ensures a common informational platform for advocacy that promotes the strategic vision and mandate of the organization.

DIGITAL-ERA SPONSORS

Technology-taking sponsors publicly endorse the idea of constant change and adaptation to new ways of working. In the digital era, leaders must also sponsor the adoption of technologies that best support value creation. Sponsors ensure that people match up with processes and that both people and processes change to meet the demands of technology.

Another characteristic of an effective sponsor is that she is willing to sustain her leadership and engagement over the course of the change project.[117] Yet the change in the digital-era is never-ending; so too, a sponsor's role. Sponsorship cannot be an add-on to an executive's to-do list. Instead, being a sponsor for digital-era technology change becomes part of the executive's very job description. If the organization's leader was unaware she had signed up for this permanent work and sustained commitment, she may be unable to discharge her sponsorship duties effectively.

As organizations adopt more and more complex, digital-era technologies, sponsors may struggle to understand how these technologies operate and the depth of organizational behavior change they will cause. This challenge is partly a function of the fast-moving, modern age: as soon as we grasped Facebook, everyone jumped to Instagram; having mastered that, they moved onto Snapchat; we had just downloaded that app when they stampeded ahead to KiK. Technologies are being developed and implemented at the seeming speed of light.

The digital era's sponsorship difficulty also may be a function of the sponsors themselves. It takes time to assume the leadership, power, and authority necessary to be an effective sponsor. The average age of a Fortune 500 CEO is now 58 years old, having grown older, not younger, the recent tech boom notwithstanding.[118] Warren Buffett (who is now 85 years old) famously did not invest in the technology sector because he did not understand the science behind it.[119] Only recently, after the iPhone wave had crested and well after the dominance of desktop computers had settled, did Buffet invest in Apple and IBM stock. Unlike most Buffett investments, these have lost value.[120] Although the organization's interns may well be familiar with and adept at using the latest technologies, the people in the C suite may not.

Change Organizational Culture

Shortcomings in organizational culture are one of the main barriers to company success in the digital era.[121] Managers will fight the digital-era changes imposed on them, not wanting technology to determine processes or behavior. Managers will argue that the organization's every system must be customized to reflect in-house terminology, language, and steps. Sponsors will have to resist these arguments.

Digital-era sponsors ultimately are responsible for changing the organization's culture from technology resistant to adaptable. They further understand the full scope and human impact of the change and the nature of its likely resistance. With their strong belief that the proposed change is the right solution for their organization, sponsors are willing to sacrifice personally and to dedicate organizational resources sufficient to achieve the change.[122]

Sponsors engage their organizations in a vision for a digital culture that involves being proactive, customer-centric, data-driven, agile, risk-taking, based on test findings, and cross-functional.[123] A technology-taking culture seeks ways to innovate and removes unnecessary layers to speed up decision-making, break down silos, and improve communication.[124]

Sponsoring Actions for the Technology Taker

Sponsorship is a worthwhile investment, as without it the organization surely will fail in the digital era. The sponsorship endeavor is no longer for one or more projects but a continuous effort of adaptation, so the payoffs for executive sponsorship have also changed. A leader that does not act adeptly with appropriate investment of their political capital in the sponsorship of the digital era is not playing to win at technology taking.

Several sponsoring actions are critical for the technology-accepting change leader: (1) state frequently and publicly the commitment to using technology as a driver of behavior change, (2) set out a directional mandate for change, thereby managing technology taking's effect on people, (3) suspend use of outdated processes and require global good practices and (4) lead by ensuring there are is a space for change advocacy within the organization that can result in a multiplied effect across all sponsored change management actions.

Verbalize the Commitment to Technology-taking

In the digital era, technologies create constant change. Sponsors cannot describe all changes to come; they can help their organizations anticipate change and grow comfortable with its constancy. Sponsors are asking that others trust in the value that digital-era technology will bring to their organization: In Machinis Fidimus.

The purpose of sponsorship is to change the way people think, not solely how they behave. Technology taking requires the use of language as generative of behavior change.[125] The way change leaders talk about something affects how they think and act. Sponsors can explain what it means to be a technology taker, and sponsors can engage their organizations in dialog about technology-taking concepts. What people speak, they believe. What they believe, they can do.[126]

Many organizations look at change sponsorship communications as a one-way model of informing people of decisions after the fact: "We have decided what we are going to do. So now we will inform you so that you will know what to expect." In some cases, organizations pretend to use a two-way model of communication: "We have decided what we want to achieve, and we would like to engage you in determining how to do it." Both of these communication models for sponsoring technology as a driver of behavior change are flawed; they assume a sure ability to describe what is changing and condescendingly assume that all important decisions are made at the top.

Sponsors must constantly voice aloud that change is on the horizon. Once is not enough. Twice is not enough. Three times is not enough. Once sponsors feel they are talking about the technology-taker initiative three times more than they need to, managers will begin to believe that the sponsor is personally committed.[127]

The director of the SaaS ERP-adopting UN agency publicly declared her sponsorship of this new technology.[128] She held fourteen dialogs with all people to discuss the discontent and the excitement of being technology takers.[129] The conversations often were hard and uncomfortable, especially when people were informed that technology would determine the agency's work processes and would demand new skills from workers.[130] It was presumed, though, that grievances should be said aloud and responded to, instead of being left to fester. In every dialog, the director insisted there was a great benefit to all the change: The organization would be able to reduce costs, meet its mission, and thrive in the new era.[131]

Mandate Behavior Changes

As one of its acts of sponsorship, an organization's top leadership develop a mandate that says why, what, and how everyone in the organization needs to contribute to the desired change. In the digital era, a change mandate focuses on both change direction and technology's effect on people. Through their mandate, sponsors make clear that all organizational people will change their workplace behaviors to respond to new technologies, and that these technologies will improve working conditions by making them more efficient and effective.

A mandate formally gives notice that the organization will have a new change direction by adopting a technology and adapting to that technology's requirements. No one in the organization can opt out of these changes. Without a written change mandate, people can view adoption of new technologies as having a limited effect and as able to be ignored.

A mandate states that technology is no longer subsumed to people and processes. It makes clear that the digital era requires technology-taking, and the organization cannot stand apart from the technological forces moving the world. The mandate explains that the new technology adopted will determine new processes used, and not vice versa. And mandates will direct people to change their work practices and to find new ways of working that involve using new, technologically determined processes. Finally, a mandate can define the parameters against which success will be judged, and to which people will be held accountable for results.

A mandate for change must also convey a clear direction: that being a technology taker is everyone's responsibility for the sake of supporting the organization's mission and, if nothing else, avoid its disintermediation. Staying as a technology tailor is to remain as the calvary, which was ultimately disbanded as a force in modern warfare. As their last act, perhaps leading to a final success as at the Battle of Schoenfeld, Tailors might charge ahead blind to the digital era. But the war ultimately will be lost with use of old methods. For success in the digital era, everyone in an organization needs to become a technology taker. A digital-era technology taker is able to make decisions or suggest them, even if these challenge business-as-usual, to move forward the organization's adaptation to technologies.

Suspend Use of Outdated Processes and Require Global Best Practices

Digital-era technologies are driving new ways of conducting business, and these new processes are the same for all who use the technologies. Sponsors should redirect their organizations from over-investment in internally defined processes and to acceptance of processes determined by new technologies. New ways of working are inevitable, and the old ways are becoming obsolete.

One sponsor ensured that his team used SharePoint by refusing to respond to emails with attachments (rather than SharePoint links). While he had vocalized his expectations in meetings, initially people continued to create individual documents, rather than taking the additional time to post them in SharePoint. By consistently emphasizing the importance of a single source of truth and continually ignoring emails with attachments, the sponsor shifted everyone's behavior away from the outdated process.[132]

Sponsors need to focus on how digital era technologies are changing the processes that govern relationships between employers and employees and between internal and external stakeholders. But they also need to stop sponsoring the reliance on systems that were built in the 1970s and 1980s. Kodak invented digital pictures but refused to use them; Kodak has disappeared from the market. This same fate may befall big banks. COBOL, the coding language on which banking mainframe systems are based, is the domain of a dwindling niche of coders; and, the "COBOL blues" are real for the non-technology-taking banking sector.[133] A staggering 95% of ATM swipes still rely on COBOL; 43% of banks rely on the language for their main IT system.[134] Yet the COBOL-speaking "IT cowboys" are now retiring.[135] COBOL is the underlying reason why financial technology is ripe for disintermediation.

Unfortunately, banks are not alone. Too many organizations believe that their future successes are guaranteed by their current internal procedures and controls, and no plan is made to address the digital era. Sponsorship means leading the changeover from internally defined processes to technology taking.

DIGITAL-ERA ADVOCATES

For the digital era, advocacy is the role of every technology taker. To advocate is to support a cause and to convince others to do so. Advocates are able to influence others based on their personal credibility and reputation. People will believe in technology taking when someone they know and trust believes in technology taking. Advocates are often more trusted and accessible than sponsors. Advocates personally demonstrate their commitment to technology taking by their actions, becoming role models of technology taking.

Advocates engage with their community by making members feel heard and able to make a difference in the organization. The concept of advocacy has long been used in political contexts as a method of educating lawmakers and raising to their attention issues of importance to various constituencies.[136] Because advocates are viewed as accessible, an organization's rank and file feel more at ease approaching them with concerns and doubts about technology-taking. Advocates can convey these concerns to the organization's change sponsors; and, as they themselves are users of the technologies being adopted, advocates can suggest more acceptable behavior changes.

When sponsors understand the essential role of advocates, they can harness this powerful force. Advocates can go places and interact with people in ways that many sponsors cannot. They are able to demonstrate an empathy that many sponsors lack, and understand the perspectives of individuals who are not managers. The best advocates are able to work across organizational boundaries as they further the case for change.

Creating Advocacy

Advocates must possess credibility, connections, and company intelligence.[137] They are considered opinion leaders, people who are respected for their expertise and experience, but they may not have a position of power. Sponsors need to devote extra time and attention to identifying these individuals, helping them understand the need for technology taking.[138]

Sometimes an effort to reach out and treat people with respect can cultivate advocates. One major American energy producer struggled to implement a change program due to resistance from middle managers.

Top-down one-way communication was not having the desired effect. Out of desperation, the top executives held a series of one-on-one conversations with middle managers, describing the business case for change and the anticipated impact on employees. Following these conversations, the change began to happen, for the managers were converted to the cause and began to advocate for it.[139]

In order to build advocacy for their innovation initiative, Whirlpool executives sent employees drawn from different levels and functions to visit companies in Japan and Korea. Whirlpool leaders anticipated that first-hand observation of competitors would convince these employees that the planned changes were both appropriate and feasible. The plan worked. Once convinced, the employees became advocates who shared their experience and conclusions with their peers. Whirlpool empowered and mobilized these advocates by convening and videotaping peer discussion groups led by them.[140] Then, these advocacy videos were used to attract still more supporters for change.

ADVOCACY ACTIONS FOR THE TECHNOLOGY TAKER

For the technology taker ready and willing to advocate for digital-era change, three actions are recommended: (1) develop a strong network in all functions and geographies, (2) model the desired behaviors and (3) reinforce the adaptation within your sphere of influence.

Develop a Strong Network in All Functions and Geographies

As described, one organization successfully identified advocates in each department and office as a support for launching its cloud-based ERP system. While not every advocate embraced the role, those that did formed a strong cadre of catalysts for change within their spheres of influence.[141] The community of advocates is built through the shared language, philosophy, and experience of technology taking.

Creating a new digital era vocabulary helps people break with the past, embrace the new, and develop a sense of community. On a mid-1990s trip to Seattle, we decided to try a new coffee shop with a mermaid logo. We asked for a medium-sized cup of coffee; but, surprise, surprise, Starbucks does not offer anything in "medium." Our choices were *tall*, *grande*, or *venti*. To become a caffeine taker at Starbucks required the acquisition of a new language.

Technology takers are not always able to retain and speak their native language. The medium may change to *Grande*. A technology taker might differentiate between "phone" and "smartphone" or iPhone and Android. These distinctions in language affect the way people think. While this is difficult at first, in the long run, technology takers are able to communicate and share the experience with others who understand their new digital-era language, enabling additional productivity and sparking new ideas. Around this new language and viewpoint, a community is formed.

Once a community of advocates forms around the mutual appreciation for the language, philosophy, experience, and strategy of technology-taking, this community must be sustained and strengthened. The theory of Legitimate Peripheral Participation suggests the more individuals within a community interact, the more they learn, and the better the community becomes at addressing the issue it seeks to address – technology adoption and adaptation or any other challenge.[142] Digital systems can facilitate interaction, collaboration, and skills/knowledge exchange across an advocacy community, allowing newcomers to benefit from improved access to advice, support, and resources from the identified "expert" advocates.[143]

Model the Desired Behaviors

Advocates strengthen the bonds of their community using the very digital-era technologies whose adoption they advocate. It is no accident that one's Facebook contacts are called "friends." Communicating and sharing content and digital experiences with others via Facebook build a common understanding among people who may not ever meet IRL (in real life). Relationships are not at arm's length or merely virtual; they are seen as friendships.

The International Society for Technology in Education (ISTE) is an organization that supports teachers who advocate for the use of technology to empower all students as learners. ISTE teachers learn, meet, and interact to become leaders and advocates for student equity and access to technology. ISTE advocates practice what they preach, engaging with one another via Twitter Chats, Voxer groups, and other technology-enhanced learning activities.[144]

Research from the Republic of Ireland and Northern Ireland points to the efficacy of using digital-era tools of engagement to build advocacy communities and to bring about changes in beliefs and behavior. As part of a campaign for women's healthcare, advocates designed and used an "exploratory protosite" to engender conversations among stakeholders. A study of this effort found that digital storytelling can help reject false narratives and raise awareness of realities. The researchers suggest the proactive use of contemporary technology to curate narratives that "provoke empathy, foster polyvocality, and ultimately expand the engaged community."[145] Furthermore, this research calls upon technology makers to support advocacy through conscious design of their technologies to aid activism.[146] Such technologies and technological applications could include social media sharing features or email, text, and Twitter templates for sharing concerns.[147]

Reinforce Adaptation Within the Sphere of Influence

The advocates lead technology-taking at all levels of the organization, wherever the advocate has influence. If the advocate interacts with five people in the left corner of the third floor, then that is where she models using digital-era technologies in her own work. If the advocate has an opportunity to advocate the sponsor, all the better. She can strengthen the sponsor's resolve to support ongoing change and may improve the sponsor's understanding of digital-era technologies.

Advocates use lobbying techniques to gain support for their cause.[148] When faced with a detractor, an advocate determines what motivates that person and what of value could be traded for that person's support. One author worked with an organization that was implementing a total quality management program. She could not understand why one well-respected employee was resisting everything, and causing others to follow his lead. Finally, someone clued her in. This particular employee had written a book on total quality management. When the change team bought everyone copies of his book and asked him to lead workshops, he changed from an antagonist to supporter overnight.

Advocates also try to build general support for particular issues or changes, as the real enemy of change is apathy.[149] Much of advocacy is creating excitement around an issue. As the train starts to move, and changes start to happen, more and more people jump aboard because they fear being left behind. Advocates can use their communities and the very technologies they champion to increase the conversation and awareness of technology taking.[150] Adoption and adaptation of technology should be positioned as modern and what "everyone" is doing.

The ultimate goal of advocacy for technology-taking is to create a supportive, positive environment for adopting new technologies and a willingness to undertake behavior change. Advocates are challenged to reach those nonbelievers who have resisted their organization's change management efforts. Advocates cannot succeed solely talking to those converted to the technology-taking cause. Instead, they must expand their spheres of influence and seek out those not yet reached by other communication efforts. Changing people's minds occurs through meaningful dialog, whether in-person or via virtual means, about the difficulties the digital era poses and the promises it might bring.

ENGAGING ORGANIZATIONS IN TECHNOLOGY-TAKING

The Behavior Change Delta at Chapter 1 reminds that the digital era requires placing technology before people and processes. Sponsors of technology engage their organizations with a technology-taking strategy that affects both people and processes. Technology-taking sponsors realize that using digital-era technologies will require behavior changes that will put the organization in a better position to create value. Sponsors, therefore, mandate that all people in the organization must change. But digital-era sponsors also realize that technology-taking is not just sponsorship at the top; change will occur if supported by a distributed leadership model of adaptation and adoption. A model of change where engagement through mandated sponsorship and advocacy creates interaction effects.

Toyota looks beyond sponsors for change leadership. *Nemawashi* is the first step in the decision-making process within the Toyota Production System.[151] The term is drawn from Japanese gardening and describes a special technique of uncovering the roots before transplanting trees. Each portion of the root system is given individual attention in order to prepare it for the impending change.[152] In the corporate context, nemawashi means engaging each part of the organization in the dialog to address their concerns.[153]

Toyota understands that dictates from top management do not generate the personal commitment required for sustained organizational change. Sponsors must exercise their power. However, that is not enough. In order for change to take root, individuals must be honored, heard, and actively engaged.

We call for change agents or champions to take on a new role as advocates. Organizations must take the time to identify, nurture, and empower their advocates. Once organizations involve them in technology taking efforts, advocates can demonstrate to their colleagues that relying on the primacy of digital-era technology will create a hopeful, and not fearful, future.

PLAY 4: EQUIP PEOPLE WITH THE SKILLS OF THE FUTURE

> **The Play:** *Invest so people are equipped with technology skills for the future.*

AT&T is spending a billion US dollars to retrain its entire workforce of 250,000 people.[154] The company found that it had "no choice" but to undertake this training re-haul; the digital era had rendered obsolete most of its employees' technology skills; and forecasts that by 2020, 75% of its network will be controlled by software-defined architecture. That percentage was almost zero in 2000.[155] The new digital-era landscape requires skills in cloud-based computing, coding, data science, and other technical capabilities. These fields are growing and changing at such a rapid pace that it cannot be presumed that organizations or their people can adapt to them without investment.

To meet this challenge, AT&T required that the company and its employees co-invest in strengthening its workforce. The company provides employees tools to identify their own skills deficiencies and to prognosticate future workforce needs. Then, employees must also spend their own time and money to fill these "skills gaps." AT&T, Udacity, and Georgia Tech have collaborated to develop online courses, certifications, and degree programs for employees, who spend five to 10 hours a week completing them.[156] From January to May 2016, retrained employees filled half of all technology management jobs at the company and received 47% of promotions in AT&T's technology department.[157]

This play describes addressing the change management needs of people in the Behavior Change Delta. To implement a technology taker strategy, organizations need to support their people's obtaining digital-era skills. The 2017 United Nations System Leadership Framework, now being implemented throughout the UN, suggests that factors pulling and keeping qualified, forward-thinking people into organizations include comprehensive training on an organization's digital systems and continuous opportunities for learning about the use of digital systems.[158] These will make attractive an organization's approach and the behavior changes required to build a modern organizational culture.

INVEST IN PEOPLE TO EQUIP THEM FOR THE DIGITAL ERA

Extensive investment in skills is needed for technology taking to be successful. Yet, some purveyors of the digital-era technology argue that their technologies are so intuitive that training is not required for successful adoption. One technology maker's sales pitch touts its "on demand and on the go" training modules — available on social media and video; no in-person training is envisaged as necessary by this technology maker.[159] A project organizer technology trumpets, "No Training Required: [Our] easy-to-use interface takes no time to learn, and is incredibly flexible."[160]

While it is true that using Google, the leading Internet search provider, involves no training, most digital-era technologies represent a significant change in work practices and mindset. Although workers can self-learn how to use a particular technology, people cannot by themselves revolutionize how they do their jobs and how to move from doing tasks to managing information. For example, adopting standard ERP software affects the working style of an organization and requires business process change.[161] Education helps ensure desired behavioral changes in the acceptance and use of information technologies.[162]

A review of a technology that changes organizational communication from email to hashtags and channels points out that adoption of new, easy-to-use technologies may decrease efficiency. Without training, workers may become overwhelmed, especially if they "lack maturity" to prioritize tasks or suffer from a "fear from missing out" that causes them to play with the technology instead of getting work done.[163]

Organizations that go with a zero-training approach make technology taking harder than it needs to be. The sales pitches of technology makers are tantalizing, as they dangle before organizations the cost-saving prospect of no training necessary. But organizations will reap quickly diminishing returns from penny-wise and pound-foolish decisions not to support their personnel's transition to the digital era.

Unisys Corp. learned this lesson the hard way during a companywide rollout of Windows XP and Microsoft Office 2003. After dropping new software applications on employees and skimping on training about the use of these applications, Unisys ended up spending a lot of money after the fact. Workers needed hands-on guidance about how to apply the new technology tools to the jobs they had. Since that experience, Unisys has taken a much more proactive approach to training. Similar to the change advocates described in Play 3, Unisys utilizes early adopters of software to identify troubles spots and help create customized training exercises.[164]

Organizations too are misled into believing their people will not need training because digital-era technology has user-friendly interfaces. Ability to use a technology is distinct from understanding how to use it. Workers will require instruction in the technology itself and also in the replacement of old ways with new ways of working, the interconnection of processes and people, the analysis of data, and the decision-making from data-drawn conclusions.

Workers, therefore, must be trained in the philosophy and actions of technology taking. Willingness to change behavior will be based on personal understanding of why the organization must transform itself for the digital era. Organizations must explain their decision to implement a technology taker strategy. Organizations can point out that it is now the twenty-first century and that they are trying to avoid disintermediation. They can also point to other, similarly situated organizations that are using the same technologies. Comparison against other organizations indicates how specific business processes might be used, the business transformed, and success defined.

It takes time to invest and make the transition to a technology taker approach. McDonald's transition to customers' self-ordering via in-store kiosks or smartphones required training and investment in people. Workers had to be repositioned from taking orders to serving tables.[165] Other successful companies too have invested in engagement, communication, and training on how to become technology takers. Domino's, a large pizza restaurant chain in the United States, now considers itself a delivery company based on digital-era technology, rather than exclusively a pizza maker.[166] Fully half of the 800 employees at Domino's headquarters work in software and analytics to support the company's planned pizza deliveries by drones and robots.[167]

There are abundant ways to model the use of technologies and to share knowledge with those willing to receive it, from classroom training to peer-to-peer modalities. Learning management systems can monitor with ease of knowledge transfer activities. The key strategic decision that technology taking organizations must make, as AT&T did, is to budget to support employee learning and training initiatives.

TRAIN ON MORE THAN TECHNOLOGY USE

Investment in training places people at the center of the change management effort, even if the technology is driving the need for change. Senior management must sponsor training, and the CMF should coordinate it. Training is aimed at equipping people with the knowledge and skills necessary for the digital era, not only teaching new tasks.

The governance play reminds leaders that they must ensure training for all audiences, from workers to middle managers to executive leadership, who require it. Oftentimes, change business cases, policies, and procedures are presented and explained to managers – but what about the salesforce, administration, assistants, and secretaries? Not providing training in a coordinated fashion to all people in an organization risks incomplete inculcation about the organization's norms, the technologies to be adopted, and the processes to use them.

When investing in training, organizations should prioritize investment in building skills that will equip the technology taker to act with resolve in the face of digital-era dislocations. Effective training changes mindsets and habits, updates ideas, and corrects misapprehensions about technology adoption and adaptation. Play 4, Equip, teaches people what they must know about and builds acceptance of the digital era.

Explain the Reasons for the Change

Becoming a technology taker requires significant change management action, and the people component of the Behavior Change Delta requires special attention and investment. People require an explanation of why and to what end the organization is adapting to the digital era. Employees must understand and accept the reasons for the change before they are willing to participate, learn, and advocate for them. "People need to understand that change is required and be given reasons why they need to change."[168]

As described in Play 3, sponsorship is paramount to learning and eventual behavior change. Unfortunately, occasionally sponsors fail to communicate why people are being asked to change. The authors worked with a government agency who failed to demonstrate strong sponsorship for change. The agency trumpeted the need for people to act, then sent to training classes long-time civil service employees who were unprepared mentally or emotionally to learn a new way of working. After about an hour, the trainers left crying. The trainers were unable to answer the employees' question, "Why do we need to do our work differently?" Training can answer "how" questions; sponsors must answer "why" questions prior to acting with conviction on setting the virtuous cycle of change in motion.

Internal training too is used to explain participation in the governance of technology-taking, including innovations to policies and procedures. Training is an opportunity to distinguish between policy and process, discuss what procedures could be furthered by technology-taking, and identify what should be altered to improve value creation. Training about governance also explains and documents how and by whom policies and procedures could be changed.

Relate to an Organization's Mission

Change management and training need to be part of an overall strategy that is aligned with the organization's mission. Play 1 described how the CMF must develop a business case for change. This case will include continuing and organization-wide efforts to deliver technological adoption and adaptation, and the CMF will ensure these delivery efforts include change management interventions and training. Yet people will not accept a different way of working, regardless of the amount of training offered, if they do not understand how their actions support the organization's mission. The relationship between technology-taking and mission cannot be assumed to be intuitive.

For its retraining initiative, AT&T required change management to continue to meet the expectations of its stakeholders regarding its mission to be "a world leader in communications, media and entertainment, and technology."[169] To maintain its position as a world leader, AT&T as a company needed to adopt and adapt to those technologies used by other leading companies. AT&T's people had to contribute to this effort by developing top-of-the-class skills and understanding of the digital era. Neither the company nor its workers could rest upon the laurels of their past successes, as these matter little in the modern world. Nor could AT&T or its employees use past strategies of ignoring technology or customizing technologies, as neither technology tinkering nor tailoring now results in world leadership.

Build a New Mindset

Sometimes leaders spend months or even years mentally and emotionally determining the course they would like to see for a major organizational change. When decisions have finally been made to opt for technology-taking as a strategy, leaders may continue to downplay and devalue the time they spent getting personally prepared. Since most organizations have little appetite for focus groups or workshops to surface issues, the major – and indispensable – forum for workers' mental preparedness for change is training.

In the context of a major change, training is not just about teaching new skills or procedures. Training is often leaders as well as individuals' first experience that informs them how much they are being asked to change. For technology takers, most digital-era technologies represent a big change in reflexive work practices. The end-user interface may be intuitive and easy to learn, but people do not change their ways of doing things quickly. Facebook users do not automatically also adopt Instagram; many still keep their digital camera even though their smartphone now takes better pictures; and people will still flag a cab rather than use Uber. Everyone will need to adapt, and the initial training classes are the moment this fully sinks in as a new mindset.

Astute leaders will use training courses as an opportunity for themselves to learn and to communicate, to acknowledge people's frustrations, and invite people to act by participating in the creation of their own future. Training will describe and model desired behaviors in accordance with the specific, mapped out changes of the Behavior Change Delta. Training should map out what is going to change, where, and for whom.

Workplace change evokes fear, as employees foresee potential losses: loss of status, loss of expert knowledge, loss of control, and, often, loss of a job. The newer and more far-reaching the technology, the more likely that employees will fear layoffs. As the story goes, the factory of the future will have two employees, a man, and a dog. The man will be there to feed the dog. The dog will be there to keep the man from touching the equipment.[170]

Whether or not they are warranted, such fears can cause a visceral stress response: fight or flee.[171] People experiencing these fears are unable to exploit new technologies, as they are preoccupied with finding ways to hide or get out. Or workers revert to unproductive tinkering or tailoring of technology instead of accepting the necessary conditions of technology taking.

Training can help address fears by reducing uncertainties and building confidence. Sponsors can use training sessions to allay fears, coax people out of hiding, and encourage them to stay with the change. Sponsors draw attention to the potential value that can be created through the acquisition of new skills and renewed mindsets.

Imbue New Habits

According to Phillippa Lally, a health psychology researcher at University College London, it takes more than two months on average for a new behavior to become a habit.[172] Some people take more than eight months. It is the height of folly to assume that a one-day class will enable people to act in a way that signifies a change to their work habits – even when they are motivated to do so. Fully embracing new technology often requires a mindset change. Humans can rewire their brains, but doing so takes more than sitting through a sponsorship presentation. It takes practice. "The more something is practiced, the more connections are changed and made to include all elements of the experience (sensory info, movement, cognitive patterns)."[173]

Successful technology-taking organizations offer training participants the opportunity to practice in the new workplace environment. Internal experts can give appropriate context and answer questions about how work will be performed. This fieldwork approach can "hardwire" training into capability-building processes for individual job roles and new skills.[174] Employees may attend courses at a variety of venues alongside people from other organizations in order to learn the latest technologies. Companies need to stay up to speed on digital technologies that might enhance their work, and continually ask "what if we adopted this technology?"

Employees also require ongoing training on the technologies that have been adopted by their organization as part of technology-taking. External specialists and internal subject matter experts must demonstrate "how to apply this technology." Ongoing internal training programs are required to help people understand changes to their individual roles and tasks while sustaining the link to their works continued contribution to the organization's mission.

As people learn why the organization is moving to technology-taking, they begin to understand exactly how they will perform their work tasks in the future. They build new habits more quickly. The more tangible and tactile the training experience, the more people can move forward mentally and see the opportunities, rather than only the dangers. Change management in this respect is a practical exercise that helps drive new habits to support technology taking.

Convey a New Skills Philosophy

The philosophy of technology taking must also be made explicit to managers and people as part of the change management effort. This will shift the organization from tinkering or tailoring or even dithering, waiting for the disintermediation to come.

The 2016 American film *Arrival* – where a linguist acquires a new concept of time after deciphering alien conversations – is based on the Whorfian hypothesis that language can influence mindset.[175] By learning the technology maker's language, a technology taker will be able to think in different ways. This cognitive shift occurs slowly; but once the learner grasps the overall mental structure behind a particular technology, including why technology making may not be possible, tinkering unprofitable and tailoring cost prohibitive, suddenly understanding accelerates. This requires repeated exposure and learning experiences that build not only new habits but reinforces them through the conveyance of a new skills philosophy.

When they implemented new ERP systems, four United Nations agencies with a global-scope in both their missions and population of end-users deemed intensive user training "critically important" to their change management efforts.[176] These agencies found some training of higher priority than others. Even more pressing than explanations of specific ways to navigate applications were up-font efforts for sponsors to explain why the new systems had been selected and how to gain acceptance of the ERP systems' underlying philosophies.[177] In particular, the International Atomic Energy Agency (IAEA) expended significant effort assisting its people to understand general data analysis concepts and data structures.[178] Managing using data streams is a fundamental change from the enabling era and demonstrates how a new skills philosophy is required for both adaptation and adoption to digital-era technologies.

DIGITAL-ERA SKILLS ENUMERATED

Many people, like those employees at AT&T, grew up in an era where life was lived in three sequential phases: learn, work, retire.[179] In the traditional model, people focused on learning until their early twenties, at which point they got a job and applied what they learned. Today, many people are working in jobs that did not exist when they were in school. The skills they learned in school may well have become irrelevant. To keep up, everyone from senior management to employees needs to invest in lifelong learning.[180] Without a commitment to lifetime learning, people and organizations cannot successfully navigate to the digital era.[181]

With digital-era technology, executives need to know how to access, interpret, and utilize information. These skills must be placed within the context of management tasks of analysis and decision-making. The following seem to us good metrics for evaluation of digital literacy:

- Photo-visual literacy: ability to read and interpret visual representations of data.

- Reproduction literacy: ability to creatively recycle existing material.

- Branching literacy: hypermedia and nonlinear thinking.

- Information literacy: evaluating information with a skeptical eye.

- Socioemotional literacy: willing to share data and knowledge with others, capable of information evaluation and abstract thinking, and able to collaboratively construct knowledge.[182]

Of these five metrics of digital era literacy skills, the most important is the socioemotional, or the willingness to learn. In most organizations we have worked with, leaders and managers especially do not demonstrate a willingness to learn something new. Many managers participate in continuing education in their field; however, they often do not want to appear as novices in a new area. Managers' reflexive attitude is that it is better not to attempt something, especially related to technology, than to look foolish in front of peers and subordinates.

Invest in Digital-era Skills

In order to encourage others to learn about new technologies, leaders and managers visibly need to demonstrate their own willingness to learn technologies. A manager who understands digital-era technology is able to apply it productively do her work, recognizes when it would assist or impede the achievement of a goal, and adapts continually to the changes and advancement in technology to be deemed Fluent with Information Technology (FIT).[183] FITness requires a more complete mastery of technology for information processing, communication, and problem-solving than does mere "computer literacy."[184]

One of the challenges leaders face in setting up people with FIT skills for them to act on technology-taking is that learning digital technology often relies on instructional design targeting technologists, those who already have a technology skillset. Instructional design recognizes that business leaders rarely volunteer to learn wiring or coding or network protocols, as this would be analogous to requiring new drivers to attend a car mechanic course. Rather, it would be more helpful for a leader or manager to know how to drive safely and how to navigate from point A to point B than to understand how to refurbish an engine.

However, a Tesla is rarely chauffeured; to drive one requires using the complex dashboard screen. Functioning in the digital era requires basic technical knowledge. As AT&T realized, people needed to be taught digital-era skills in accessing, interpreting, and using information obtained from cloud-based, constantly changing technologies. Investments had to be made to explain new data flows, new approval processes, new ways of analyzing and reporting data, and new approaches to decision-making.

One of the oldest and largest healthcare staffing firms in the United States, CHG Healthcare Services, has consistently been recognized for its corporate training program.[185] What started as a specialized niche, technical training has grown into a critical building block of corporate training. CHG provides technical training to help employees learn the technical components of their jobs, including business standards such as Microsoft Excel and company-specific applications.

People can be technology takers only if they are trained on how to extract and analyze data, as well as how to present resulting analysis in a meaningful way to aid decision-making. Data flow is extremely difficult to conceptualize. People need the opportunity to interact with information that is meaningful. This means that investment needs to go into creating training data that can be manipulated so people can see how changes are going to affect them. A demonstration illustrating new roles and approvals or role play and hands-on experiences can support what might otherwise feel like a lack of control.[186]

Ultimately the investment in digital era skills is about the need for a data-driven opportunity for faster decision-making, turning technology takers into critical thinkers. Technology taking change management is about facilitating critical, profitable thinking based on the objective analysis of facts to form a judgment.[187]

To establish a minimum of modern technical knowledge in its people, the UN Development Programme (UNDP) created its International Computer Driving License initiative.[188] This effort served as the basis for additional training modules on business processes, change management, e-documentation, and enterprise resource planning. For 70% of the participants, these sessions were their first-ever management training at UNDP.[189] Effective learning for adults of the UNDP type is self-directed, utilizes knowledge and life experiences, is goal-oriented and relevancy-oriented, highlights practicality, and encourages collaboration.[190]

The successful training uses many different techniques in a variety of fora that are accessible to the people who need it. AT&T offers employee short and long courses and professional degree programs. The very first training experience for any new technology can be overwhelming, no matter how wonderful the materials and instruction. Participants may be inundated with new concepts, terms, and tasks. One way of reducing the initial overwhelm is to use internal experts as teachers within a classroom setting. When people hear from others they know and respect, they are more likely to believe that the technology taker approach will actually work in their environment.

Training can occur through informal means, such as reading groups or learning circles. In these, a group of people meets to discuss books or articles relevant to the organization. Meetings usually take place outside normal working hours, such as lunch time.

The organization should provide materials to support training efforts. Compiling a comprehensive dictionary of new vocabulary may be helpful. Although companies may not always find a one-to-one relationship between old terms and new terms, the dictionary should offer a crosswalk from the old language to new language. Or newspaper articles, government announcements, and reports can be used as learning alerts. These can be summarized for employee consideration and included on the agenda of staff meeting for a brief discussion.

Finally, attention should be paid to the knowledge measurement techniques, which will be further discussed in Play 5. Training assignments should have quantifiable, outcome-based measures that indicate levels of competence gained.[191] Providing people certification recognizes and rewards the skills attained.[192]

BUILD THE SKILLED WORKFORCE OF THE FUTURE

Surviving in the digital era is not easy. Disintermediation lurks around every corner and is far from your control.[193] What to do? Equip people with the skills they will require, not only in using technology. Invest to ameliorate people's skills gaps in what would be required for faster movement toward the technology taker quadrant.

Organizations equip their people mentally and practically to improve their technology skills and change their behavior. They undertake this investment not out of altruism, but to be able to respond to constant digital era disruption. Cultivating modern skills and mindsets means the organization's people better support its mission.

With its billion-dollar price tag, AT&T's investment in its workforce comes with the expectation that AT&T's managers will successfully re-invent themselves as technology takers. The effectiveness of AT&T's Workforce 2020 retraining effort is measured in four categories – activity, hydraulics, business outcomes, and sentiment.[194] Within these categories, particular metrics of managerial competency include digital literacy in technologies needed for the company's future (in *Activity)*, willingness to take on new roles (in *Hydraulics*), and increases in efficiency (in *Business Outcomes*).[195] AT&T's redesigned compensation plans de-emphasize seniority, add more variable compensation to motivate high performers, and give weight to the in-demand technology skills. The company offers financial rewards for individuals with skills in high demand, including cybersecurity, computer science, data science, IT networking, and software-defined networking.[196]

AT&T made a strategic decision to retrain its workforce instead of replacing it wholesale. But the writing is on the wall for AT&T's managers: they have until 2020 to become FIT for the digital-era future, and AT&T will provide them the training and tools to become technology takers. Managers and people must reinvent themselves for the digital era, or they must get out. To give everyone a chance to stay FIT, organizations should equip their personnel with the skills of the future.

PLAY 5: MEASURE MANAGERS' EMBRACE OF TECHNOLOGY CHANGE

> **The Play:** *Measure managers' shift from merely observing the digital era to embracing and participating in change.*

With apologies to Charles Dickens, we present a tale of two committees with their wisdom and foolishness.[197] We worked with two, high-level, managerial committees within two multinational organizations. Each committee took a distinct approach to preparing for meetings, conducting meetings, and communicating decisions made during meetings. One organization had digitally savvy managers, while the other organization had digitally remiss managers. Which organization was the technology taker and which the technology tinker could be easily observed.

One committee's table was covered in paper, yet needed documents were often waylaid. Presentations were made of lengthy, dense reports, which consumed the time of each manager to slog through prior to determining next steps for action. A secretary wrote down decisions made by others; but, when the notes were distributed, the decisions were usually in dispute.

The other committee convened around a paper-free table. Each member pulled out her laptop to peruse the bulleted action items emailed to them prior to the meeting. When tasks were assigned, each member noted down her planned activities in a collaborative working document displayed before the entire group on a large screen. All were in agreement about assignments and decisions made, and meeting notes were stored on-line, should memories have required refreshing (Table 1).

The digitally savvy managers are technology takers. They invest the time to identify and adopt digital-era tools. They use commonly available hardware and software. They have modified their own behavior and processes to improve personal efficiency and effectiveness. And they set an example to be followed for everyone working in their organization.

The committees demonstrate ways that managers will change their use of and attitudes about technology. A technological embrace will assist the performance of the fundamental management tasks of learning

Table 1: A Tale of Two Committees.

Digitally Savvy–Single Source of Data	Digitally Remiss–Paper Everywhere
• "Don't bring paper, email me ahead of time"	• " This 50-page report is too short, needs to be thicker"
• Meetings conducted using large computer screen with no handouts	• Meeting table is covered with lots of scattered documents
• Participants bring their own laptops to look up any details	• Participants bring subordinates to provide details
• One set of notes compiled and distributed electronically	• Everyone takes notes and argues later who is right

information, making decisions using information, and clearly communicating those decisions. These actions build managers' accountability for the speed and effectiveness of decision-making and for related organizational prosperity.

MEASURE EMBRACING CHANGE

According to Lord Kelvin (for whom the Kelvin scale of absolute thermodynamic temperature was named), "If you can measure it, you can manage it; If you cannot measure it, you cannot manage it."[198] For leaders to know if technology taking has set down roots in their organizations, they must measure managers' embrace of technology change. Many managers are happy to pretend that they are responding to their organizations' goals to move to the digital era, but they continue to work in the same old ways. Managers must be measured by their willingness to change.

Leaders must be both technology takers and digital-era change managers. All in an organization need to understand that technology is today non-negotiable. Leaders and managers, including legal, finance, and executive officers who might think of themselves exempt from technology should be tested on technology adoption and awareness. If their technological abilities are not measured and reviewed regularly, some mistakenly will believe they can opt-out of being technology takers. They will not be equipped to lead digital era change management and embrace technology solutions essential to the survival of their organizations in the digital era.

The business effect of a refusal to become technology takers was illustrated when one manager took a taxi to a business meeting in NYC. Staying at a boutique hotel in Brooklyn, the manager jotted down an address from a laptop screen and handed a slip of paper to the smiling concierge. The concierge called the taxi dispatch service and relayed the address provided. Except, a small mistake was made in the analog passing of information. The mistake was discovered when one passenger cried out, "Aren't we crossing the Brooklyn Bridge!?" The desired destination was in Brooklyn, but the taxi drove to Manhattan. The manager was late to the meeting, and the bill for the whole excursion was at least 20% above what it should have been.

Ride-sharing apps have a feature for the user to type in the destination address and to verify it visually on a map. Using digital-era technology to hail a ride reduces human error. Savings are found, and value created, in blue dollars (people time) and green dollars (direct costs). The concierge is disintermediated, as is the taxi dispatcher.

There may be many good reasons for not using ride-sharing technology, but the principle of the person closest to the information inputting and managing that information has real cost implications. Entering one's own information into a technological application, creating a clear data stream when doing so, and managing that data stream analytically is not only a matter of organizational creation of value. It is also an essential component of digital-era change management and change leadership. Managers failure to embrace technology impedes an organization from measuring its performance. A lack of data streams means that results cannot be compared with those from other technology users, wasting an opportunity provided by cloud-based technology.

HOLD MANAGEMENT ACCOUNTABLE

Measuring is for many organizations still a tale of the emperor's new clothes. Hans Christian Andersen had the insight that people can so fear authority, they are indisposed to call out deceptive management behavior. That is, until one child famously outs a naked emperor by shouting, "[H]e isn't wearing anything at all."[199] Measurement of managers' attitudes about and abilities to use modern technology will reveal who is properly attired for the digital era.

When leadership does not hold itself and its managers accountable for using technology, it is likely the technology will go unused and behaviors unchanged. "[M]ost executives don't see themselves as 'part of the problem' and therefore deep down do not believe that it is they who need to change, even though in principle they agree that leaders must role model the desired changes."[200]

General Electrics (GE) needed to hire outside help to conduct an internal review to understand why a spare, empty, private jet accompanied its chief executive on his travels — and why the CEO was unaware that he was being shadowed by an empty airplane.[201] There was no monitoring using data or other performance metrics set up relative to the instructions of the GE board for senior management. Or if there were, then the person closest to the information, the CEO, did not embrace technology to own that data and set an example to his people on what was and was not appropriate. No one dared call out the emperor on lack of technology taking for what was and is an organization purporting to be at the forefront of the digital era.

Accountability for one's own travel arrangements only scratches the surface of the level of behavior change required by senior managers. Change is needed if managers are to be effectively measured against metrics that show the technology taking of their organizations. An idiosyncratic or privileged approach should not be an alternative.

When executives adopt new digital technologies, they can improve the way they conduct business. Their personal example will drive similar efficiencies down the chain of command, with more impact than any memorandum or training class. When leaders clearly demonstrate that they are technology takers, that they are willing to make personal sacrifices and change their own behavior in order to exploit digital technologies, their employees take notice. Excuses for avoiding technology taking will fall on deaf ears.

The aim of technology-taking managers should be to participate in and lead the virtuous cycle of change management. Managers should evaluate their own behavior, ensuring they do not make exceptions for themselves. All managers need to change behavior and become role models in their organization. Doing so provides practical, highly visible encouragement for technology-taking at all levels. Managers need to show that they too are being measured by driving the Behavior Change Delta of their organization.

Leaders are responsible for ensuring the measurement of their managers. As "emperor" of their organization, a leader cannot risk being outed as naked and powerless for the digital era. Leaders must own, and not delegate, their responsibility for technology taking and for the "weaving" of technology into their organizations.

MEASURE DIGITAL-ERA COMPETENCY

Technology competency has to be a feature of twenty-first-century management.[202] Leaders need to know about the technologies that catalyze renovation of work practices and behaviors, leading to change in the organizations under their management. Yet technological ability is not often measured as a core competency of modern managers, even if formal competency frameworks may say otherwise.[203]

Those who abrogate their responsibility to embrace technology become digitally remiss, are not technology accountable, and fail to fully understand and support what their organizations require to transition and move through the dislocations of the digital era. The difference between Facebook and MySpace demonstrates this point. Facebook's executives were quick to respond to changing user preferences and technology, resulting in the company's success. MySpace was acquired by NewsCorps whose executives did not have the proper competencies to understand the MySpace technology or business and whose decisions were bogged down in a quicksand of a "bureaucratic morass."[204] The company is described as a "technology failure."[205]

IBM's process reengineering efforts revealed that success was ultimately dependent on including seasoned, technology competent managers in process teams.[206] Managers must demonstrate that they have the necessary skills to compete in the digital-era marketplace. The technology skill of management will determine the speed with which an organization can move northeast to technology taking within the Adoption–Adaptation Strategy Matrix.

Way back in 1995, Bob Martin, the CIO turned CEO of Wal-Mart Stores' International Division, observed that a global company's ability to take information from multiple systems and make it broadly accessible to managers and employees was critical to his organization's success.[207] Walmart required managers to reexamine their technology knowledge. Those who could not use technology to analyze and synthesize data could not manage effectively.[208]

Over 20 years later, the stakes have been raised. Startup companies that exploit digital technologies can put established companies out of business overnight. Digital savvy is not "nice to have" with respect to business decisions. All executives need to regard understanding and embracing digital-era technology as part of their competency-DNA and business decision-making framework.[209]

A fundamental concept of the digital era is that processes are externally dictated by the technology chosen. As a business decision, instead of developing their own supportive, auxiliary processes, companies have found it more efficient and economical to use "third-party products and services – digital Lego blocks."[210] Managers must be savvy in the sense of having awareness of the existence of these offerings down to the capabilities they may provide for their organizations' transition to the digital era.

Registered investment advisers, which have been the fastest-growing segment of the US investment-advisory business, purchase record keeping and operating infrastructure from Charles Schwab, Fidelity, and others.[211] These "turnkey" systems from technology makers give individuals or small groups all they require to run their own firms as technology takers. Managers are left to concentrate on their market advantage: the provision of investment advice.[212] To do so, however, managers need to know that technologies exist for their taking and they need to find ways to measure how they are doing on using that technology.

"Technology has redrawn the lines of what businesses can do, what is expected and what job roles typically entail [...] failure to understand and embrace technology as part of leadership responsibility is a failure to ensure profitability."[213] Competent managers use digital-era technology and measure their change leadership as a way to reimagine possibilities for their businesses, thereby generating value. Executives can no longer view technology through the lens of their existing tailored processes or their reengineering, asking, "How can we use these new technological capabilities to improve what we are already doing?"[214] Instead, modern managers should be asking, "How can we use technology to allow us to do things that we are not already doing?"[215]

CONNECTING DIGITAL-ERA LITERACY TO INDIVIDUAL PERFORMANCE

Digital-era technologies aggregate data from multiple users, permitting organizations to compare themselves and their people with others using the same technology. Turning the challenge of measurement into a benefit, using digital measurements immediately demonstrates which people are using a digital-era technology, and which are not. Benchmarks further make managers aware of improvements that are orders of magnitude beyond what they would have thought possible, and they shake managers and employees out of complacency.[216] Organizations should factor the results of benchmarking measurement into criteria for promotion of managers.[217]

To connect organizational change management to digital literacy and management performance at the level of the individual manager is a tall order. Tools that offer benchmarking and performance management options for people will appear as highly invasive of both personal and managerial privacy.[218] These tools are themselves products of the digital-era, collecting data on each worker's efficiency, effort, and results for retention decisions.[219]

Conveniently, the digital era provides behavior changing options to solve performance management problems. Connecting digital literacy to individual performance can, for example, be achieved through the use of data as a service (DaaS). DaaS allows the comparison across technology takers of the same underlying processes. Sharing data across various technology takers of the same process affords the ability to gain an understanding of how an organization compares to other organizations.[220] Organizations can even be in different industries.

Benchmark comparison need not stop at the organizational level. It can be extended to units, functions, and individuals, including managers. Individuals are already being benchmarked by ride-sharing drivers as users of the transport service through their star system.[221] Connecting hard data with technology literacy and individual performance is part of a broader trend in the on-demand economy.[222] It redistributes managerial oversight and power away from formalized management and toward a triadic relationship among employers, workers, and consumers.[223] All aspects of this relationship can be measured.

Data-driven individual performance monitoring will become more pervasive as interactions of all sorts are tracked electronically and compared through analysis of data.[224] Using data streams for performance can raise ethical issues about how they are created and used. These discussions are a corollary of already ongoing appropriateness debates about data-driven performance management accountability.[225]

Organizational leadership will increasingly use data streams for change management purposes. Those purposes include connecting digital literacy to individual performance to root out digitally re-missive managers in favor of digitally savvy managers. Those who have acquired the competencies of the digital era will be best positioned to survive in a benchmarked world that will increasingly measure performance down to the individual level. Data streams will permit comparison across organizations rather than subjective, qualitative performance reviews. Where individual performance shows that the embrace of technology is not possible, especially at the leadership level, separation may be required.

Prepare to be Measured

Irrespective of digital benchmarking options, performance management as a discipline is undergoing significant change. The complexity of demands for performance management continues to be a feature of most organizations, with some linking individual performance objectives to ever more complex result-frameworks.[226] Alternatively, some seek to move away from classic batch evaluations at mid-year and year-end.[227] The classic performance review is turned on its head, and leaders are asked what they would do with their team members, not what they think of them.[228]

For the digital era, a renewed focus on competency frameworks and job descriptions is a more fruitful avenue for how managers can prepare to be measured by their digital literacy. First, job descriptions will need updating to match benchmarkable tasks with required digital competencies like knowledge acquired, skills demonstrated, and attitudes communicated.[229] In this model, each of the required competencies includes levels for ranking one to five, where one is an expert and five is a novice. Development plans are then added as qualitatively focused yet measurable reflections on how managerial duties are shifting as the organization becomes a technology taker. A blended performance management approach will help convert the competencies measured into personal performance action plans.

Now couple the benchmarking capabilities of DaaS with the derived transparency of organizational as well as individual performance, and the accountability for digital era literacy quickly gains relevance. There are no escaping future performance data streams, but the crux of change leadership is to play with a conscious knowledge that digital savvy and delinquency affects the ability of an organization to deliver on its mission. For this reason, the measurement of the embrace of technology by managers must be taken seriously.

For example, suppose an individual requires a level-four competency in photo-visual literacy. Her development plan might include reading (such as the classic *Visual Display of Quantitative Information* by Edward Tufte, 2001), courses, or working with a mentor in order to acquire knowledge. She could be given special assignments in creating graphics for document and presentations in order to demonstrate skills. Coaching on new attitudes, such as avoiding the use of PowerPoint for data-dense discussions,[230] could include encouragement to conduct brown-bag session or formal presentations or writing an article in the company newsletter. Lastly, seeking out benchmarks against which to measure oneself is required.

We can hear the groans. No one has time for personal development plans, and skills like "photo-visual literacy" sound like IT department skill-sets not relevant to managers. Most people have not yet experienced the transparency of live-data dashboards that relate directly to your performance or that of the organization. When these arrive, everyone will find that the benchmarking is unforgiving. If the organization and its managers do not prepare to be measured, there are repercussions beyond the creation of day-to-day value and into the ability to participate in the virtuous cycle of digital era change management. If an organization does not offer serious performance management apt for the digital era, others will.[231]

To conceptualize cutting-edge, digital-era performance management is to forward change management capabilities of an organization, both at the individual and organizational levels. The two competencies that best support change management skills are risk management and failure acceptance. Ultimately, the objective of measuring managers' embrace of technology is innovation through trial and error and the capability to push improvements that require risk taking.

Encourage Risk

The willingness to be a technology taker requires support or encouragement of appropriate risk taking for people and managers to stay tech-savvy. Managers should free up engagement and learning spaces to try out new approaches as "without risk there is no innovation."[232] In many organizations, it is a relatively easy exercise to run a survey or a set of targeted interviews to uncover if the organization is taking risks commensurate with its stated risk-taking appetite or policy. Too often, risk identification and management crowd out the task of assessing the capacity and creation of space in an organization for risk-taking. Innovation stems from risk encouragement, and in "large organizations, innovation culture tends to be dictated by senior management."[233] Leaders need to make the risk-taking baseline and stance explicit as a matter of good governance. If not, then encouragement of risk cannot take place in a managed fashion and will not anchor itself as a core competency of digital-era adaptation.[234]

Peter Robertson observed that Dutch fishermen usually fish together.[235] The entire fleet will sail to an area where fish tend to gather. There is a risk with this approach - one day the fish may move elsewhere and everyone comes home with their nets empty. In order to mitigate this risk, two boats do not fish with the fleet. They go to different waters, testing to see if schools of fish may have relocated. It is this constant testing that keeps all the fishermen in business.

Organizations can establish performance evaluation systems to encourage managers to be explicit and practical about risk management, like the Dutch fishermen. Connecting risk tolerance and the fishermen's mitigation strategy, one way to gain information about managerial performance on hard-to-measure metrics (such as willingness to adopt the technology) is to seek feedback from professionals outside of the organization.[236] The entrepreneur Steve Andriole has spoken of his company's establishment of an Innovation Lab that includes participants from outside his company. As there is no innovation without risk, the purpose of including these outsiders is to push his company's managers to "pursue open innovation that leverages the ideas of the best minds regardless of where we find them."[237]

Accept Failure

Technology takers are innovators and risk takers, and organizations that value their managers' risk-taking and experimentation will have a better chance of success. However, not every attempt to use new technologies works out as planned. Organizations should not penalize managers who make a good faith effort to adopt a new technology - but fail. In fact, failure is imperative to organizational learning and may have significant long-term benefits.[238]

For example, a smartphone user who takes a photo of a barcode in order to compare product prices has no idea that bar coding was not designed with retail products in mind. Bar codes were originally affixed to a different kind of stock to solve a different kind of inventory problem. Large bar codes were used on rolling stock − train cars − to keep track of their location in train yards that spanned miles. They did not work well, since dirt accumulated on the bar codes and made them difficult to read accurately. The Universal Product Codes (UPCs) of today owe their origins to a failed technology for transportation.[239] It is not a stretch to consider how the Internet of things and edge-collected data will continue to appear as a failure to the uninitiated; yet it is likely a question of time before failure succeeds.[240]

Organizations could regard managers' failures as evidence of their achievement of the metrics of technology taking. Managers who fail in technology adoption nevertheless were willing to take on new roles and ways of working. And they evaluated information and demonstrated abstract thinking. International conglomerates, such as the Tata Group, even award failures by repackaging bold attempts as "daring to try."[241]

For managers not to be penalized for technology-taking failures, organizations must be sure to review their performance according to their willingness to share knowledge with others and to construct responses collaboratively. The CMF and related governance of change management play a key role in documenting both success and failure. Managers who fail must re-evaluate and be helped by the CMF to do so, asking why success did not occur and sharing these lessons throughout the organization. Managers also must review failure patterns, to assure themselves that lessons for change are being synthesized and understood.[242] Knowledge management rightly understood can provide unbiased coverage of a chain of events, all of them: the good, the bad, and the ugly.[243] By doing so, what is generated and supported as an organizational competency is useful knowledge because, "while we as individuals and organizations can learn from what went right, we are also inclined to learn from mistakes and failures."[244]

EMBRACE TECHNOLOGY CHANGE

Technology-taking is a savvy change strategy that, if applied correctly, offers leaders and organizations additional choices for a competitive edge. In recent years, the travel industry has been reinvented and reinvented again, as the market moved away from brick-and-mortar agencies to online do-it-yourself planners to virtual, niche providers of complete travel packages. Technology-taking small firms and individuals offer consumers entire trips, including flights, hotels, and car rentals, from online portals that assemble this information in real time, with dynamic pricing that depends on supply and demand.[245] Capitalizing on these widely available technologies, trip providers do not spend their time developing websites or purchasing air tickets. Instead, they compete to craft bespoke travel "experiences" made specifically for each client.

Reinvention as done by technology-taking travel agents is never easy. Leaders and their organizations can learn to embrace technology change. The embrace of technology can now be measured by everyone who cares to look. Observation makes it clear who are the digitally savvy. Data-powered benchmarking reveals those managers most adept at using a technology's processes. The CMF can guide managers' technological adaptation, using a business case approach to technology reviews or by briefing senior management and change managers on current technologies.

Managers rarely want to look less learned in front of peers and subordinates. A low starting point of knowledge can lead to fear; fear can lead to questions not being asked. Lack of questions, especially at senior levels, can heighten digital delinquency even as the digital era charges on.

A way to ease fear is to create engagement spaces where information can flow freely. Peer advisory boards are conducive to open discussion of new ideas. For example, the United Nations System Staff College uses a peer-network as foundational to support senior political leadership under its United Nations Laboratory for Organizational Change and Knowledge (UNLOCK), the purpose of which is "to share knowledge and foster collaboration for change."[246] Discussion groups can be safe spaces for managers to admit lack of knowledge and to learn from others,[247] especially when conducted under Chatham House rules, which protect the identity of participants.[248]

Organizations may further seek to reduce digital delinquency by developing a relationship with a technology mentor. A community of other technology takers will provide encouragement to one another. Access to experts can be found in technology advisory organizations that collaborate on knowledge sharing, both commercial as well as noncommercial knowledge incubators, universities, and other learning institutions.

Behavior change to become a technology taker can be supported by measuring managers' performance and the results of their efforts vis-a-vis technology.[249] Managers should be evaluated against specific behaviors related to the desired competencies that will drive business performance in the digital era.[250] The rationale behind establishing metrics of managerial success is that what gets measured has a higher propensity to get done.[251]

Organizational technology taking can be assured only if managers themselves become technology takers.[252] Leading-edge organizations and their leaders will have to embrace technology and continually improve their technology savviness rather than remain remiss. With this knowledge and these capabilities, managers can participate actively in the virtuous cycle of digital-era change management. Technology-taking managers will contribute to their organization's successful move toward the northeast quadrant of the Adoption–Adaptation Strategy Matrix.

PLAYBOOK COMPLETED: THE NEXT FRONTIER

Digital technologies and digital innovation are no longer something "new."[253] The playbook for digital-era change leadership makes clear that whatever is next matters for technology. After Play 5, change leaders need to complete the Virtuous Cycle of Change and return to Play 1 to re-envision their continuous change management for the digital era. A new vision anticipates future changes that must be managed.

New technologies will change the technology taker strategy and the plays that the technology taker needs. The next frontier of technology is unknown: "change may be inevitable but technological advances not only inevitably result in further change, but also change the nature of change."[254] What is clear is that implementation of the technology taker strategy is contingent on change management.

The playbook to digital-era change leadership helps organizations revisit and renew their understanding of what it means to compete in the digital era. The nature of and approach to technology-driven change and change management are more malleable than project or risk management. The latter two disciplines appear to have settled on solid ground in recent years. In contrast, change management is still a profession in which the mere mention of the term puzzles people. We have experienced organizations where no activity could be labeled change management because of previous poor experiences. Due to poor experiences, some leaders see change management as a problem, not a solution.[255]

For technology, the area of change most in flux is artificial intelligence (AI), which likely will have a significant effect on the people-side of change.[256] AI is devoted to making machines intelligent, and intelligence is that quality that enables an entity to function appropriately and with foresight in its environment.[257] AI's functioning will challenge change leaders to respond not only to the people but also to the machines' determinations of what is "appropriate" and whether foresight is apt to new, unpredictable challenges.[258]

Change management innovation, including as a response to technology taking, is a thought process on how to help leaders catch up with reality, a reality of rapid dislocations through disintermediation and general digital-era upheaval. BMW, the German automobile manufacturer, has created an ethics team to help guide development of automated or self-driving cars.[259] Presumably, BMW has done so because technology can have unintended ethical consequences, regardless of how well it has been change managed. How a technology taker can stay within a virtuous cycle of change must be carefully considered.

With this warning, do not stop now, at the end of the playbook. Instead, cycle back to make sure that there are contingencies in place for re-envisioning technology taking, renovating the plays as tools, and applying the techniques. As technology redefines management,[260] what is changing and will need change managing will continue to require attention and redevelopment.

NOTES

1. Messina, C. (May 30, 2017). *The UN leadership framework: A catalyst for culture change at the UN.* Retrieved from http://www.unssc.org/news-and-insights/blog/un-leadership-framework-catalyst-culture-change-un/

2. Focuses on business transformation by drawing on a range of edited work, including case studies and the authors experience from working with McKinsey: Swaminathan, A., & Meffert, J. (2017). *Digital @ scale: The playbook you need to transform your company.* Hoboken, NJ: Wiley.

3. Lewin, K. (1935) *A dynamic theory of personality.* New York, NY: McGraw-Hill. Lewin, K. (1936) *Principles of topological psychology.* New York, NY: McGraw-Hill.

4. Bridges, W. (2017). *Managing transitions: Making the most of change.* Da Capo.

5. Conner, D. (2006). *Managing at the speed of change: How resilient managers succeed and prosper where others fail.* New York, NY: Random House.

6. Kotter, J. P., Abrahamson, E., Kegan, R., Lahey, L., Beer, M., Nohria, N. ... Linsky, M. (August 25, 2015). *Leading change: Why transformation efforts fail.* Retrieved from https://hbr.org/2007/01/leading-change-why-transformation-efforts-fail

7. Roberto, D. A. (February, 2005). *Change through persuasion.* Retrieved from https://hbr.org/2005/02/change-through-persuasion

8. Ewenstein, B., Smith, W., & Sologar, A. (July, 2015). *Changing change management*. Retrieved from https://www.mckinsey.com/featured-insights/leadership/changing-change-management

9. Everett, C. (January 19, 2017). *How culture change has to underpin success in digital transformation*. Retrieved from https://www.computer-weekly.com/feature/How-culture-change-has-to-underpin-success-in-digital-transformation

10. *Harvard Business Review*. (November 03, 2016). *The four phases of project management*. Retrieved from https://hbr.org/2016/11/the-four-phases-of-project-management

11. Kotter, J. P. (July 13, 2015). *Leading change: Why transformation efforts fail*. Retrieved from https://hbr.org/1995/05/leading-change-why-transformation-efforts-fail-2

12. Groysberg, B. et al. *The leader's guide to corporate culture*. Retrieved from https://hbr.org/2018/01/the-culture-factor

13. Groysberg, B. et al. *The leader's guide to corporate culture*. Retrieved from https://hbr.org/2018/01/the-culture-factor

14. *Our purpose*. (February 17, 2016). Retrieved from https://www.virgin.com/virgingroup/content/our-purpose-0

15. *Our culture | virgin atlantic careers*. (n.d.). Retrieved from https://careersuk.virgin-atlantic.com/life-at-virgin-atlantic/culture

16. Perry, D. (November 20, 2014). *Sex and Uber's 'rides of glory': The company tracks your one-night stands – and much more*. Retrieved from http://www.oregonlive.com/today/index.ssf/2014/11/sex_the_single_girl_and_ubers.html

17. Ibid.

18. Likierman, A. (October 01, 2009). *The five traps of performance measurement*. Retrieved from https://hbr.org/2009/10/the-five-traps-of-performance-measurement

19. Kotter, J. P. (July 13, 2015). *Leading change: Why transformation efforts fail*. Retrieved from https://hbr.org/1995/05/leading-change-why-transformation-efforts-fail-2

20. Messina, C. (May 30, 2017). *The UN leadership framework: A catalyst for culture change at the UN.* Retrieved from http://www.unssc.org/ news-and-insights/blog/un-leadership-framework-catalyst-culture-change-un/

21. Kupersmith, K., Mulvey, P., & McGoey, K. (n.d.). *How to write a cost/benefit analysis for a business case.* Retrieved from https://www.dummies.com/business/business-strategy/how-to-write-a-costbenefit-analysis-for-a-business-case/

22. Kouzes, J. M., & Posner, B. (January, 2009). *To lead, create a shared vision.* Retrieved from https://hbr.org/2009/01/to-lead-create-a-shared-vision

23. Levin, M. (March 30, 2017). *Why great leaders (like Richard Branson) inspire instead of motivate.* Retrieved from https://www.inc.com/ marissa-levin/why-great-leaders-like-richard-branson-inspire-instead-of-motivate.html

24. Lawson, E., & Price, C. (June, 2003). *The psychology of change management.* Retrieved from https://www.mckinsey.com/business-functions/organization/our-insights/the-psychology-of-change-management

25. Ibid.

26. Ibid.

27. Ibid.

28. Ibid.

29. Morris, P. W. (February, 2011). *Brief history of project management.* Retrieved from http://www.oxfordhandbooks.com/view/ 10.1093/oxfordhb/9780199563142.001.0001/oxfordhb-9780199563142-e-2

30. Elton, J., & Roe, J. (March, 1998). *Bringing discipline to project management.* Retrieved from https://hbr.org/1998/03/bringing-discipline-to-project-management

31. International Organization for Standardization. (November, 2009). *Risk management – Principles and guidelines.* Retrieved from https:// www.iso.org/standard/43170.html

32. Mind Content Tools Team. (n.d.). *Porter's five forces: Understanding competitive forces to maximize profitability*. Retrieved from https://www.mindtools.com/pages/article/newTMC_08.htm

33. *UN blockchain: Multi-UN agency platform*. Retrieved from https://un-blockchain.org/category/wfp/

34. Ibid.

35. World Food Programme. (n.d.). *Building blocks*. Retrieved from http://innovation.wfp.org/project/building-blocks

36. *Why a business case is key to your digital transformation | digital transformation*. (n.d.). Retrieved from https://www.panorama-consulting.com/why-a-business-case-is-key-to-your-digital-transformation/

37. Ibid.

38. Jain, A., & Beale, A. (2017). Developing a business case for digital investments in health and social care. *International Journal of Integrated Care, 17*(5). doi:10.5334/ijic.3633

39. *Calgary-born Robert Opp leading UN World Food Programme's new innovation division*. (July 14, 2006). Retrieved from https://betakit.com/calgary-born-robert-opp-leading-un-world-food-programmes-new-innovation-division/

40. Lawrynuik, S. (January 03, 2018). Albertan born on a grain farm to rethink how World Food Programme's humanitarian aid is delivered. *CBC News*. Retrieved from http://www.cbc.ca/news/canada/calgary/robert-opp-world-food-progamme-innovation-alberta-1.4471461

41. Prosci. (n.d.). *A change management office primer*. Retrieved from https://www.prosci.com/change-management/thought-leadership-library/a-change-management-office-primer

42. Creasey, T. (n.d.). *Latest data and key considerations for the CMO*. Retrieved from http://blog.prosci.com/Latest-Data-and-Key-Considerations-for-the-CMO

43. The Risk Management Society, RIMS, was convened in 1950 to serve the insurance industry and, though its Enterprise Risk Management Center of Excellence, broadened its scope in 2005 to meet the more

expanded needs of corporate risk management. See "RIMS Annual Report 2005." Retrieved from https://www.rims.org/aboutRIMS/AnnualReports/Documents/2005annualreport.pdf

44. The Institute of Risk Management was founded in 1986 and launched its International Certificate in Risk Management in 2005. See *Our story*. (n.d.). Retrieved from https://www.theirm.org/about/our-story.aspx

45. International Organization for Standardization. (November, 2009). *Risk management – Principles and guidelines*. Retrieved from https://www.iso.org/standard/43170.html

46. Louisot, J., & Ketcham, C. H. (2014). *ERM enterprise risk management: Issues and cases*. Chichester: Wiley.

47. Christensen, J. *The decision to internally generate or outsource risk management activities*. August 2011. Retrieved from https://epublications.bond.edu.au/cgi/viewcontent.cgi?article=1095&context=theses

48. Based on authors' consulting experience.

49. Desmet, D., Löffler, M., & Weinberg, A. (n.d.). *Modernizing IT for a digital era*. Retrieved from https://www.mckinsey.com/business-functions/digital-mckinsey/our-insights/modernizing-it-for-a-digital-era

50. *UN blockchain: Multi-UN agency platform*. (April 04, 2018). Retrieved from https://un-blockchain.org/category/wfp/

51. Milano, A. (April 20, 2018). *€2 million donation to fund world food programme blockchain project*. Retrieved from https://www.coindesk.com/world-food-programme-blockchain-project-receives-e2-million-donation/

52. Ibid.

53. Ibid. See also Alexandre, A. (April 12, 2018). *Belgium contributes to world food programme blockchain project*. Retrieved from https://cointelegraph.com/news/belgium-contributes-to-world-food-programme-blockchain-project

54. Tirone, J. (February 16, 2018). *Banks replaced with blockchain at international food program*. Retrieved from https://www.bloomberg.com/

news/articles/2018-02-16/banks-replaced-with-blockchain-at-international-food-program

55. Ibid.

56. *UN blockchain: Multi-UN agency platform*. Retrieved from https://un-blockchain.org/category/wfp/

57. Kotter, J. P. (July 13, 2015). *Leading change: Why transformation efforts fail*. Retrieved from https://hbr.org/1995/05/leading-change-why-transformation-efforts-fail-2. See also, Conner, D. (August 15, 2012). *The real story of the burning platform (describing the burning platform approach to creating a sense of urgency)*.

58. Kotter, J. P. (July 13, 2015). *Leading change: Why transformation efforts fail*. Retrieved from https://hbr.org/1995/05/leading-change-why-transformation-efforts-fail-2

59. Ibid.

60. Hsu, T., & Kang, C. (March 26, 2018). *Demands grow for Facebook to explain its privacy policies*. Retrieved from https://www.nytimes.com/2018/03/26/technology/ftc-facebook-investigation-cambridge-analytica.html

61. Rosenberg, M., Confessore, N., & Cadwalladr, C. (March 17, 2018). *How trump consultants exploited the Facebook data of millions*. Retrieved from https://www.nytimes.com/2018/03/17/us/politics/cambridge-analytica-trump-campaign.html

62. *A cloud-based ERP renovates work practices and changes behavior at PAHO* (Case Study Series, pp. 1-15.). United Nations System Staff College. Retrieved from http://www.unssc.org/sites/unssc.org/files/mini_case_study_unssc_02_fin.pdf

63. This play does not examine corporate governance, that is, board and shareholder responsibilities of publicly traded corporations, as set out in the Sarbanes-Oxley Act of 2002 (116 Stat. 745) or the Model Business Corporations Act (2008).

64. Vantrappen, H., & Wirtz, F. (December 26, 2017). *When to decentralize decision making, and when not to*. Retrieved from

https://hbr.org/2017/12/when-to-decentralize-decision-making-and-when-not-to

65. Dunleavy, P. et al. (2006) *Digital era governance* (p. 83).

66. *A cloud-based ERP renovates work practices and changes behavior at PAHO* (Case Study Series, pp. 1–15.). United Nations System Staff College. Retrieved from http://www.unssc.org/sites/unssc.org/files/mini_case_study_unssc_02_fin.pdf

67. Wakabayashi, D., & Shane, S. (June 01, 2018). *Google will not renew pentagon contract that upset employees*. Retrieved from https://www.nytimes.com/2018/06/01/technology/google-pentagon-project-maven.html

68. Heath, A. (June 22, 2017). *Facebook has a new mission statement: 'to bring the world closer together'*. Retrieved from http://www.businessinsider.com/new-facebook-mission-statement-2017-6

69. *Facebook and the meaning of share ownership*. (September 30, 2017). Retrieved from https://www.economist.com/business/2017/09/30/facebook-and-the-meaning-of-share-ownership

70. Wagner, K. (March 22, 2018). *Mark Zuckerberg says he's 'fundamentally uncomfortable' making content decisions for Facebook*. Retrieved from https://www.recode.net/2018/3/22/17150772/mark-zuckerberg-facebook-content-policy-guidelines-hate-free-speech

71. Schneider, N. (March 28, 2018). *Mark Zuckerberg: Give up Facebook control*. Retrieved from https://www.corpgov.net/2018/03/mark-zuckerberg-give-up-facebook-control/

72. Ibid.

73. *United Nations system staff college, delivering successful change with enterprise resource planning (ERP) systems*, Case Study Series (1/2017), p. 11 (describing ERP adoption at UNDP).

74. Stollman, J. (March 23, 2017). *Disruption: The new frontier for governance and risk professionals*. Retrieved from https://www.governanceinstitute.com.au/news-media/blog/2017/mar/disruption-the-new-frontier-for-governance-and-risk-professionals/

75. Wong, J. C. (March 22, 2018). *Mark Zuckerberg apologises for Facebook's 'mistakes' over Cambridge Analytica*. Retrieved from https://www.theguardian.com/technology/2018/mar/21/mark-zuckerberg-response-facebook-cambridge-analytica

76. Rosenberg, M., Confessore, N., & Cadwalladr, C. (March 17, 2018). *How trump consultants exploited the Facebook data of millions*. Retrieved from https://www.nytimes.com/2018/03/17/us/politics/cambridge-analytica-trump-campaign.html

77. Wong, J. C. (March 22, 2018). *Mark Zuckerberg apologises for Facebook's 'mistakes' over Cambridge Analytica*. Retrieved from https://www.theguardian.com/technology/2018/mar/21/mark-zuckerberg-response-facebook-cambridge-analytica

78. Kim, K., Jung, S., Hwang, J., & Hong, A. (2017). A dynamic framework for analyzing technology standardisation using network analysis and game theory. *Technology Analysis & Strategic Management, 30*(5), 540–555. doi:10.1080/09537325.2017.1340639

79. Günther, W. A., Mehrizi, M. H., Huysman, M., & Feldberg, F. (2017). Debating big data: A literature review on realizing value from big data. *The Journal of Strategic Information Systems, 26*(3), 191–209. doi:10.1016/j.jsis.2017.07.003

80. Davies, H. (December 11, 2015). *Ted Cruz campaign using firm that harvested data on millions of unwitting Facebook users*. Retrieved from https://www.theguardian.com/us-news/2015/dec/11/senator-ted-cruz-president-campaign-facebook-user-data

81. See Zuckerberg, *Facebook post* on Mar 21, 2018.

82. *Mark Zuckerberg, Facebook post at 12:36 pm*, Mar 21, 2018 as reposted by Sheryl Sandberg, Facebook at 12:40 pm, Mar 21, 2018. Retrieved from https://www.facebook.com/sheryl/posts/10160055807270177

83. Hern, A. (November 19, 2014). *Uber investigates top executive after journalist's privacy was breached*. Retrieved from https://www.theguardian.com/technology/2014/nov/19/uber-investigates-top-executive-after-journalists-privacy-was-breached

84. *Uber crunches user data to determine where the most 'one-night stands' come from.* (November 18, 2014). Retrieved from https://sanfran-cisco.cbslocal.com/2014/11/18/uber-crunches-user-data-to-determine-where-the-most-one-night-stands-come-from/

85. Hern, A. (November 19, 2014*). Uber investigates top executive after journalist's privacy was breached.* Retrieved from https://www.theguar-dian.com/technology/2014/nov/19/uber-investigates-top-executive-after-journalists-privacy-was-breached

86. The General Data Protection Regulation (GDPR), Regulation (EU) 2016/679, which took effect on May 25, 2018, strengthens EU citizens' rights to control their on-line data and required companies processing data of EU citizens to revise their organizations' data protection poli-cies. Most digital-era technology makers, and some technology takers and tailors, collect EU citizens' personal data in the course of doing business; therefore, the GDPR has forced policy changes of digital-era players, no matter their location. More information can be found at the European Commission's website on Data Protection in the EU, Retrieved from https://ec.europa.eu/info/law/law-topic/data-protection/data-protection-eu_en

87. Günther, W. A., Mehrizi, M. H., Huysman, M., & Feldberg, F. (2017). Debating big data: A literature review on realizing value from big data. *The Journal of Strategic Information Systems, 26*(3), 191–209. doi:10.1016/j.jsis.2017.07.003

88. Ibid.

89. *A cloud-based ERP renovates work practices and changes behavior at PAHO* (Case Study Series, pp. 1–15.). United Nations System Staff College. Retrieved from http://www.unssc.org/sites/unssc.org/files/mini_case_study_unssc_02_fin.pdf

90. *The Twitter rules.* (n.d.). Retrieved from https://help.twitter.com/en/rules-and-policies/twitter-rules

91. Blake, M. (April 13, 2018). *On disability, Twitter is better late than never.* Retrieved from https://www.cnn.com/2018/04/13/opinions/twitter-changes-terms-on-disability-blake-opinion/index.html

92. Herger, M. (June 4, 2006). *What is a business process expert, really?* Retrieved from https://blogs.sap.com/2006/06/04/what-is-a-business-process-expert-really/

93. *A cloud-based ERP renovates work practices and changes behavior at PAHO* (Case Study Series, pp. 1–15.). United Nations System Staff College. Retrieved from http://www.unssc.org/sites/unssc.org/files/mini_case_study_unssc_02_fin.pdf

94. Pastin, M. (January 21, 2018). *The surprise ethics lesson of Wells Fargo.* Retrieved from https://www.huffingtonpost.com/mark-pastin/the-suprise-ethics-lesson_b_14041918.html

95. Ibid.

96. Nocera, J. (May 18, 2018). *Wells Fargo has shown us its contemptible values.* Retrieved from https://www.bloomberg.com/view/articles/2018-05-18/wells-fargo-has-shown-its-customers-its-true-values-joe-nocera

97. Pastin, M. (January 21, 2018). *The surprise ethics lesson of Wells Fargo.* Retrieved from https://www.huffingtonpost.com/mark-pastin/the-suprise-ethics-lesson_b_14041918.html

98. *A cloud-based ERP renovates work practices and changes behavior at PAHO* (Case Study Series, pp. 1–15.). United Nations System Staff College. Retrieved from http://www.unssc.org/sites/unssc.org/files/mini_case_study_unssc_02_fin.pdf

99. Sandberg, S. (March 21, 2018) *Facebook post at 12:40 pm.* Retrieved from https://www.facebook.com/sheryl/posts/10160055807270177

100. *What is internal audit?* (n.d.). Retrieved from https://www.iia.org.uk/about-us/what-is-internal-audit/

101. Ibid.

102. *African Mother's Health Initiative.* Retrieved from http://www.africanmothers.org/

103. See tax and financial documents for the charity at *African Mothers Health Initiative.* (n.d.). Retrieved from https://www.guidestar.org/profile/26-0423197

104. Organizing governance decision-making around an operating model that establishes an organizational structure for decisions and reporting, a distribution of responsibilities to assure accountability and authority, an infrastructure for policies, procedures, communication, and technology, and principles for organizational culture and people performance. See, for example, Deloitte Development LLC. *Developing an effective governance operating model: A guide for financial services boards and management teams - Illustrative governance operating model* (2013). Retrieved from https://www2.deloitte.com/content/dam/Deloitte/global/Documents/Financial-Services/dttl-fsi-US-FSI-Developinganeffectivegovernance-031913.pdf

105. *United Nations system staff college, delivering successful change with enterprise resource planning (ERP) systems*, Case Study Series (1/2017), p. 18 (describing ERP adoption at UNOPS).

106. *A cloud-based ERP renovates work practices and changes behavior at PAHO* (Case Study Series, pp. 1–15.). United Nations System Staff College. Retrieved from http://www.unssc.org/sites/unssc.org/files/mini_case_study_unssc_02_fin.pdf

107. *United Nations system staff college, delivering successful change with enterprise resource planning (ERP) systems*, Case Study Series (1/2017), p. 30 (describing ERP adoption at PAHO).

108. Hopkins, J. (July 10, 2017). *Millennial managers: A guide for successful management.* Retrieved from https://www.forbes.com/sites/jamie-hopkins/2017/07/08/millennials-managers-a-guide-for-successful-management/#34ec6a6e2ac3

109. Creasey, T. (n.d.). *Latest data and key considerations for the CMO.* Retrieved from http://blog.prosci.com/Latest-Data-and-Key-Considerations-for-the-CMO

110. Conner, D. (2006). *Managing at the speed of change: How resilient managers succeed and prosper where others fail.* New York, NY: Random House.

111. Ibid.

112. Creasey, T. (n.d.). *Latest data and key considerations for the CMO.* Retrieved from http://blog.prosci.com/Latest-Data-and-Key-Considerations-for-the-CMO

113. Eadicicco, L., Peckham, M., Pullen, J. P., & Fitzpatrick, A. (April 03, 2017). *TIME's 20 most successful technology failures of all time.* Retrieved from http://time.com/4704250/most-successful-technology-tech-failures-gadgets-flops-bombs-fails/

114. Alsher, P. (October 29, 2013). *The CAST of characters for implementing organizational changes.* Retrieved from https://www.imaworld-wide.com/blog/bid/189157/The-CAST-of-Characters-for-Implementing-Organizational-Changes

115. O'Connor, G. C., Corbett, A. C., & Peters, L. S. (2018). *Beyond the champion: Institutionalizing innovation through people.* Stanford, CA: Stanford Business Books, an imprint of Stanford University Press.

116. Ibid.

117. Conner, D. (2006). *Managing at the speed of change: How resilient managers succeed and prosper where others fail.* New York, NY: Random House.

118. Cava, M. D., & Jones, C. (April 22, 2016). *For older CEOs, issue is knowing when to bow out.* Retrieved from https://www.usatoday.com/story/money/business/2016/04/19/older-ceos-issue-knowing-when-bow-out/83114728/

119. Ng, S., Holm, E., & Ante, S. E. (November 15, 2011). *Buffett bets $10.7 billion in biggest Tech Foray.* Retrieved from https://www.wsj.com/articles/SB10001424052970204323904577037742077676990

120. Cohan, P. (February 15, 2018). *Apple: Warren Buffett's second big bad tech bet.* Retrieved from https://www.forbes.com/sites/petercohan/2018/02/15/apple-warren-buffetts-second-big-bad-tech-bet/#77b054e667ff

121. Goran, J., LaBerge, L., & Srinivasan, R. (n.d.). *Culture for a digital age.* Retrieved from https://www.mckinsey.com/business-functions/digital-mckinsey/our-insights/culture-for-a-digital-age

122. Conner, D. (2006). *Managing at the speed of change: How resilient managers succeed and prosper where others fail.* New York, NY: Random House.

123. Rick, T. (April 30, 2018). *Culture change is key in digital transformation.* Retrieved from https://www.torbenrick.eu/blog/culture/culture-change-is-key-in-digital-transformation

124. Ibid.

125. Ibid.

126. Brothers, C. (2005). *Language and the pursuit of happiness: A new foundation for designing your life, your relationships & your results.* Naples, FL: New Possibilities Press.

127. Jackson, H. L. (July 13, 2015). *The hard side of change management.* Retrieved from https://hbr.org/2005/10/the-hard-side-of-change-management

128. United Nations System Staff College, *Delivering successful change with enterprise resource planning (ERP) systems,* Case Study Series (1/2017), p. 30.

129. Ibid.

130. Ibid.

131. Ibid.

132. *Sink the boats and burn the bridges!* (May 23, 2017). Retrieved from http://www.expressworks.com/sharepoint-adoption/sink-the-boats-and-burn-the-bridges

133. Irrera, A. (April 10, 2017). *Banks scramble to fix old systems as IT 'cowboys' ride into sunset.* Retrieved from https://www.reuters.com/article/us-usa-banks-cobol/banks-scramble-to-fix-old-systems-as-it-cowboys-ride-into-sunset-idUSKBN17C0D8

134. Reuters Graphics. (n.d.). *COBOL blues.* Retrieved from http://fingfx.thomsonreuters.com/gfx/rngs/USA-BANKS-COBOL/010040KH18J/index.html

135. Ibid.

136. Miller, A. P., Lewis, S., & Waites, T. (May 28, 2018). *Essentials of advocacy.* Retrieved from http://www.onlinejacc.org/content/71/22/2598

137. Weinstein, P. V. (November 05, 2014). *To close a deal, find a champion.* Retrieved from https://hbr.org/2014/09/to-close-a-deal-find-a-champion

138. Valente, T. W., & Pumpuang, P. (2006). Identifying opinion leaders to promote behavior change. *Health Education & Behavior, 34*(6), 881–896. doi:10.1177/1090198106297855

139. Lattuch, F., & Seifert, A. (2014). Insights from change management consulting: Linking the hard and soft side of change with heuristics. *Management of Permanent Change*, 177–194. doi:10.1007/978-3-658-05014-6_10

140. Armenakis, A. A., Harris, S. G., & Mossholder, K. W. (1993). Creating readiness for organizational change. *Human Relations, 46*(6), 681–703. doi:10.1177/001872679304600601

141. United Nations System Staff College, *Delivering successful change with enterprise resource planning (ERP) systems*, Case Study Series (1/2017), p. 32.

142. Michie, L., Balaam, M., Mccarthy, J., Osadchiy, T., & Morrissey, K. (2018). From her story, to our story. *Proceedings of the 2018 CHI Conference on Human Factors in Computing Systems – CHI 18.* doi:10.1145/3173574.3173931

143. Ibid.

144. Trust, T. (2017). 2017 ISTE standards for educators: From teaching with technology to using technology to empower learners. *Journal of Digital Learning in Teacher Education, 34*(1), 1–3. doi:10.1080/21532974.2017.1398980

145. Michie, L., Balaam, M., Mccarthy, J., Osadchiy, T., & Morrissey, K. (2018). From her story, to our story. *Proceedings of the 2018 CHI Conference on Human Factors in Computing Systems – CHI 18.* doi:10.1145/3173574.3173931

146. Ibid.

147. Ibid.

148. Farnel, F. (1994) *Lobbying: Strategies and techniques of intervention.* Paris.

149. *Harnessing the energy of change champions.* (n.d.). Retrieved from https://www.clemmergroup.com/articles/harnessing-energy-change-champions

150. Johansson, A. (August 20, 2015). *FOMO marketing in the age of social media.* Retrieved from https://www.relevance.com/fomo-marketing-in-the-age-of-social-media/

151. Nemawashi – Toyota production system guide. (December 09, 2014). Retrieved from http://blog.toyota.co.uk/nemawashi-toyota-production-system

152. Defining Nemawashi. (n.d.). Retrieved from http://www.japanintercultural.com/en/news/default.aspx?newsID=234

153. Ibid.

154. Caminiti, S. (March 13, 2018). *AT&T's $1 billion gambit: Retraining nearly half its workforce for jobs of the future.* Retrieved from https://www.cnbc.com/2018/03/13/atts-1-billion-gambit-retraining-nearly-half-its-workforce.html

155. Ibid.

156. Benko, C. and Donovan, John. (October 07, 2016). *Inside AT&T's radical talent overhaul.* Retrieved from https://hbr.org/2016/10/atts-talent-overhaul

157. Ibid.

158. Messina, C. (May 30, 2017). *The UN leadership framework: A catalyst for culture change at the UN.* Retrieved from http://www.unssc.org/news-and-insights/blog/un-leadership-framework-catalyst-culture-change-un/

159. Workday for Financial Services [PDF]. (2017) *Workday.* Retrieved from: https://www.workday.com/content/dam/web/en-us/documents/datasheets/datasheet-workday-for-financial-services-us.pdf

160. Trello Enterprise. (n.d.). Retrieved from https://trello.com/enterprise

161. Ahadi, H. (2004). An examination of the role of organizational enablers in business process reengineering and the impact of information technology. *Information Resources Management Journal,* 17(4), 1–19. http://dx.doi.org/10.4018/irmj.2004100101; Siriginidi, S. R. (2000). Enterprise Resource Planning in Reengineering Business. Business Process Management Journal, 6(5), 376–391. http://dx.doi.org/10.1108/14637150010352390

162. Knol, W. H. C., & Stroeken, J. H. M. (2001). The diffusion and adaption of information technology in small and medium sized enterprises through IT scenarios. *Technology Analysis & Strategic Management, 13* (2), 227–246. http://dx.doi.org/10.1080/09537320123815

163. Dewnarain, G., O'Connell, D., & Gotta, M. (October 6, 2017). *SWOT: Slack, worldwide*. Retrieved from https://www.gartner.com/doc/reprints?id=1-4K5I75F&ct=171108&st=sb

164. Stackpole, B. (March 13, 2008). *Five mistakes IT groups make when training end-users*. Retrieved from https://www.cio.com/article/2436969/training/five-mistakes-it-groups-make-when-training-end-users.html

165. Peterson, H. (June 23, 2017). *McDonald's shoots down fears it is planning to replace cashiers with kiosks*. Retrieved from http://www.businessinsider.com/what-self-serve-kiosks-at-mcdonalds-mean-for-cashiers-2017-6

166. Taylor, B. (November 28, 2016). *How Domino's pizza reinvented itself*. Retrieved from https://hbr.org/2016/11/how-dominos-pizza-reinvented-itself

167. Ibid.

168. *The role of the sponsor in bringing change to a project*. (July 06, 2011). Retrieved from https://www.brighthubpm.com/change-management/39144-sponsoring-a-change-management-initiative/

169. AT&T Profile. (n.d.). Retrieved from http://about.att.com/sites/company_profile

170. Warren G. Bennis; As cited in: Fisher, Mark. (1991) *The millionaire's book of quotations*. p. 151.

171. Cannon, W. B. (1929). *Bodily changes in pain, hunger, fear and rage: An account of recent researches into the function of emotional excitement*. New York, NY: D. Appleton and Co.

172. Lally, P., Jaarsveld, C. H., Potts, H. W., & Wardle, J. (2009). How are habits formed: Modelling habit formation in the real world. *European Journal of Social Psychology, 40*(6), 998–1009. doi:10.1002/ejsp.674

173. Merzenich, M. M. (2013). *Soft-wired: How the new science of brain plasticity can change your life*. San Francisco: Parnassus.

174. Keller, S., & Aiken, C. *The inconvenient truth about change management*. McKinsey & Company. Retrieved from http://www.aascu.org/corporatepartnership/McKinseyReport2.pdf

175. Hunt, E., & Banaji, M. R. (1988). The Whorfian hypothesis revisited: A cognitive science view of linguistic and cultural effects on thought. *Indigenous Cognition: Functioning in Cultural Context*, 57–84. doi:10.1007/978-94-009-2778-0_5

176. United Nations System Staff College, *Delivering successful change with enterprise resource planning (ERP) systems*, Case Study Series (1/2017), p. 8.

177. Ibid.

178. Ibid.

179. Bolles, R. N. (1981). *Three boxes of life and how to get out of them.* California.

180. Senge, P. M. (1990). *The fifth discipline: The art and practice of the learning organization.* New York, NY: Doubleday.

181. West, D. M. (2018). *The future of work: Robots, AI, and automation.* Washington, DC: Brookings Institution Press.

182. Eshet-Alkalai, Y. (n.d.). A holistic model of thinking skills in the digital era. *Encyclopedia of Distance Learning, Second Edition*, 1088–1093. doi:10.4018/978-1-60566-198-8.ch154

183. *Being fluent with information technology.* (2000). Washington, DC: National Academy.

184. Ibid.

185. *CHG healthcare services placed among global elite corporate training programs.* (December 22, 2009). Retrieved from https://www.news-medical.net/news/20091222/CHG-Healthcare-Services-placed-among-global-elite-corporate-training-programs.aspx

186. Freifeld, L. (October 22, 2013). *L&D best practices: Technology and technical development.* Retrieved from https://trainingmag.com/content/ld-best-practices-technology-and-technical-development

187. *Defining critical thinking.* (n.d.). Retrieved from http://www.criticalthinking.org/pages/defining-critical-thinking/766

188. United Nations System Staff College, *Delivering successful change with enterprise resource planning (ERP) systems*, Case Study Series (1/2017), p. 8.

189. Ibid.

190. Knowles, M. S., F., H. I., & Swanson, R. A. (2015). *The adult learner: The definitive classic in adult education and human resource development*. Milton Park, Abingdon, Oxon: Routledge.

191. Keller, S. and Aiken, C. *The inconvenient truth about change management*. McKinsey & Company. Retrieved from http://www.aascu.org/corporatepartnership/McKinseyReport2.pdf

192. Ibid.

193. Hankin, A. (June 28, 2018). *9 companies Amazon is killing*. Retrieved from https://www.investopedia.com/news/5-companies-amazon-killing/

194. Benko, C. and Donovan, John. (October 07, 2016). *Inside AT&T's radical talent overhaul*. Retrieved from https://hbr.org/2016/10/atts-talent-overhaul

195. Ibid.

196. Ibid.

197. "It was the age of wisdom, it was the age of foolishness": Dickens, C. (1859) *A tale of two cities*.

198. Van der Zee, I. (2002). *Measuring the value of information technology*. Hershey, PA: IRM Press.

199. Anderson, H.C. (1949). *The emperor's new clothes*. Boston, MA: Houghton Mifflin Co.

200. Keller, S. and Aiken, C. *The inconvenient truth about change management*. McKinsey & Company. Retrieved from http://www.aascu.org/corporatepartnership/McKinseyReport2.pdf

201. *GE wants to know why former CEO Jeff Immelt traveled the world with a Spare Jet*. (December 13, 2017) Retrieved from http://fortune.com/2017/12/13/ge-investigation-jeff-immelt-spare-jet/

202. Nadler, D., Shaw, R. B., & Walton, A. E. (1995). *Discontinuous change: Leading organizational transformation*. San Francisco, CA: Jossey-Bass.

203. See ACHE healthcare executive 2018 competences assessment tool. *Research & Resources*. Retrieved from https://www.ache.org/newclub/resource/competencies.cfm

204. Eadicicco, L., Peckham, M., Pullen, J. P., & Fitzpatrick, A. (April 03, 2017). *TIME's 20 most successful technology failures of all time*. Retrieved from http://time.com/4704250/most-successful-technology-tech-failures-gadgets-flops-bombs-fails/

205. Ibid.

206. Grzinich, J. C., J. H. Thompson, and M. F. Sentovich. *Implementation of an integrated product development process for systems*, 1997. doi:10.1109/PICMET.1997.653448

207. Bob L. Martin, The end of delegation? Information Technology and the CEO, *Harvard Business Review* (Sept.-Oct. 1995).

208. Martin, B. L. (September/October, 1995). *The end of delegation? Information Technology and the CEO*. Retrieved from https://hbr.org/1995/09/the-end-of-delegation-information-technology-and-the-ceo

209. "It is a losing proposition for a company's leadership to view technology as separate from business": Beckley, A. M. (August 07, 2015). *How the cloud is changing the role of technology leaders*. Retrieved from https://www.wired.com/insights/2013/09/how-the-cloud-is-changing-the-role-of-technology-leaders/

210. Hirt, M., & Willmott, P. (May, 2014). *Strategic principles for competing in the digital era*. Retrieved from https://www.mckinsey.com/business-functions/strategy-and-corporate-finance/our-insights/strategic-principles-for-competing-in-the-digital-age

211. Ibid.

212. Ibid.

213. Ton, J. (March 28 2018). *It's all greek to me − How executives can learn the language of technology*. Retrieved from https://www.forbes.com/

sites/forbestechcouncil/2018/03/28/its-all-greek-to-me-how-executives-can-learn-the-language-of-technology/

214. Hammer & Champy, *Re-Engineering the corporation*, 85 (1993).

215. Ibid.

216. Eccles, R. G. (February/March, 1991). *The performance measurement Manifesto*. Retrieved from https://hbr.org/1991/01/the-performance-measurement-manifesto

217. Noting Enterprise car rentals' factoring in promotion criteria customer satisfaction data: A. Taylor. (July, 2002). *Driving customer satisfaction*. Retrieved from https://hbr.org/2002/07/driving-customer-satisfaction

218. Describing Workday's applications that track the time employees spend on tasks and permit feedback to be sent to any employee's supervisor: *Data-crunching is coming to help your boss manage your time*. (January 19, 2018). Retrieved from https://www.nytimes.com/2015/08/18/technology/data-crunching-is-coming-to-help-your-boss-manage-your-time.html

219. Detailing how Amazon monitors every aspect of a worker's performance, requiring 80 hour work weeks and penalizing workers for taking vacation leave, spending time with their families, or suffering health problems: Kantor, J., & Streitfeld, D. (August 15, 2015). *Inside Amazon: Wrestling big ideas in a bruising workplace*. Retrieved from https://www.nytimes.com/2015/08/16/technology/inside-amazon-wrestling-big-ideas-in-a-bruising-workplace.html

220. Press Release: *Workday delivers its first data-as-a-service offering with workday benchmarking*. (n.d.). Retrieved from https://www.workday.com/en-us/company/newsroom/press-releases/press-release-details.html?id=2190890

221. The Points Guy. (December 22, 2014). *Insider series: What Uber drivers know about passengers*. Retrieved from https://thepointsguy.com/2014/12/insider-series-what-uber-drivers-know-about-passengers/.

222. Rosenblat, A. (April 07, 2016). *The truth about how Uber's app manages drivers*. Retrieved from https://hbr.org/2016/04/the-truth-about-how-ubers-app-manages-drivers

223. Wang, L. (2015). When the customer is king: Employment discrimination as customer service. *SSRN Electronic Journal.* doi:10.2139/ssrn.2657758

224. Fuller, L., & Smith, V. (1991). Consumers reports: Management by customers in a changing economy. *Work, Employment and Society*, 5(1), 1–16. doi:10.1177/0950017091005001002

225. Wang, L. (2015). When the customer is king: Employment discrimination as customer service. *SSRN Electronic Journal.* doi:10.2139/ssrn.2657758

226. Birkinshaw, J., & Heywood, S. (n.d.). *Putting organizational complexity in its place.* Retrieved from https://www.mckinsey.com/business-functions/organization/our-insights/putting-organizational-complexity-in-its-place

227. Ibid.

228. Goodall, M. B., Buckingham, M., & Ashkenas, R. (November 16, 2015). *Reinventing performance management.* Retrieved from https://hbr.org/2015/04/reinventing-performance-management

229. Waychal, P. (2016). *A framework for developing innovation competencies.* 2016 ASEE Annual Conference & Exposition Proceedings. doi:10.18260/p.26321

230. Tufte, E. (September 01, 2003). *Powerpoint is evil.* Retrieved from https://www.wired.com/2003/09/ppt2/

231. Lipman, V. (January 29, 2018). *Why employee development is important, neglected and can cost you talent.* Retrieved from https://www.forbes.com/sites/victorlipman/2013/01/29/why-development-planning-is-important-neglected-and-can-cost-you-young-talent/#7513ed86f633

232. *Without risk there is no innovation » Gofore.* (December 18, 2017). Retrieved from https://gofore.com/en/without-risk-no-innovation/

233. Ibid.

234. Ibid.

235. Robertson, P. (2007). *Always change a winning team.* Times Editions.

236. Likierman, A. (October 01, 2009). *The five traps of performance measurement.* Retrieved from https://hbr.org/2009/10/the-five-traps-of-performance-measurement

237. Andriole, S. (October 08, 2015). *What C-suite executives need to know about digital strategy and emerging technologies.* Retrieved from https://www.forbes.com/sites/steveandriole/2015/10/07/analytics-iot-social-location-security-how-to-all-get-along-a-note-to-c-suiters/#dfbc646ab27a

238. Robertson, P. (2007). *Always change a winning team.* Times Editions.

239. Brown, S. A. (n.d.). *A history of the bar code.* Retrieved from https://eh.net/encyclopedia/a-history-of-the-bar-code/

240. Tower, B. 2017. *How IoT data collection and aggregation with local event processing work.* Retrieved from https://blog.equinix.com/blog/2017/11/29/how-iot-data-collection-and-aggregation-with-local-event-processing-work/

241. Haas, J. B. (May 10, 2016). *Increase your return on failure.* Retrieved from https://hbr.org/2016/05/increase-your-return-on-failure

242. Ibid.

243. Javan, J. 2018. *Share knowledge for learning, not marketing.* Retrieved from http://www.unssc.org/news-and-insights/blog/share-knowledge-learning-not-marketing/

244. Haas, J. B. (May 10, 2016). *Increase your return on failure.* Retrieved from https://hbr.org/2016/05/increase-your-return-on-failure

245. Hirt, M., & Willmott, P. (May, 2014). *Strategic principles for competing in the digital era.* Retrieved from https://www.mckinsey.com/business-functions/strategy-and-corporate-finance/our-insights/strategic-principles-for-competing-in-the-digital-age

246. *The United Nations Laboratory for Organizational Change and Knowledge (UNLOCK).* (n.d.). Retrieved from https://www.unssc.org/featured-themes/united-nations-laboratory-organizational-change-and-knowledge-unlock/

247. Rezek, M. (July 10, 2017). *The 3 types of fear that are hindering your growth as a leader.* Retrieved from https://www.inc.com/mary-rezek/overcome-3-main-reasons-people-fear-speaking-up.html

248. *Chatham House rules* described at https://www.chathamhouse.org/chatham-house-rule

249. United Nations System Staff College, *Delivering successful change with enterprise resource planning (ERP) systems*, Case Study Series (1/2017), p. 13 (finding that performance measurement was a critical success factor for change management of technology adoption in two public organizations).

250. Keller, S. and Aiken, C. *The inconvenient truth about change management*. McKinsey & Company. Retrieved from http://www.aascu.org/corporatepartnership/McKinseyReport2.pdf

251. Attributed to Peter Drucker, but there is no clear evidence that he is the origin of this quote. Retrieved from https://en.wikipedia.org/wiki/Peter_Drucker

252. Eccles, R. G. (February/March, 1991). *The performance measurement manifesto*. Retrieved from https://hbr.org/1991/01/the-performance-measurement-manifesto

253. Morrison, Ciaran (2016) *Digital transformation strategy*. Digital Health & Care Institute, Glasgow. Retrieved from https://strathprints.strath.ac.uk/64342/

254. Pharoah, Marc. (April 9, 2018). *Transforming change management with artificial intelligence (AI)*. Retrieved from https://www.andchange.com/transforming-change-management-artificial-intelligence-ai/

255. "A legacy of failed change presents a significant and ever-present backdrop that all future changes will encounter": Creasey, T. (n.d.). *The costs & risks of poorly managed change*. Retrieved from http://blog.prosci.com/the-costs-risks-of-poorly-managed-change

256. Pharoah, Marc. (April 9, 2018). *Transforming change management with artificial intelligence (AI)*. Retrieved from https://www.andchange.com/transforming-change-management-artificial-intelligence-ai/

257. Nilsson, N. J. (2010). *The quest for artificial intelligence: A history of ideas and achievements*. Cambridge: Cambridge University Press.

258. Murphy, W. (August 09, 2017). *The nomenclature of artificial intelligence*. Retrieved from https://www.intuitiveaccountant.com/people-and-business/the-nomenclature-of-artificial-intelligence/#.W0ecD9JKhPY

259. Blons, E. (February 17, 2018). *Change management in the era of artificial intelligence: Building trust through transparency and accountability*. Retrieved from https://www.linkedin.com/pulse/change-management-era-artificial-intelligence-building-blons/

260. Kolbjornsrud, V., Amico, R., & Thomas, R. J. (November 2, 2016). *How artificial intelligence will redefine management*. Retrieved from https://hbr.org/2016/11/how-artificial-intelligence-will-redefine-management

CHAPTER 5

LEADING CHANGE IN
THE DIGITAL ERA

This book focuses on the organizational change management challenges of adopting and adapting to digital-era technologies. It explains in straight-forward terms what it means to be a twenty-first-century technology taker and change leader.

Technology taking is analogous to price taking in economics. There is no alternative to the marketplace for small firms. The market sets the prices firms may receive and the conditions in which they operate. Similarly, no other option exists than the digital era with its ever-updating and interconnected technologies. These technologies have the potential to disintermediate processes, people, and organizations while creating new and more efficient ways of working. These technologies also exact a certain price from users: to play in the digital era, one must change workplace behaviors.

Organizations must use continuous change management to take control of how modern, externally created technologies, like SaaS, blockchain, or Internet search, affect their organizations. As defined for the digital era, change management leads an organization to adopt, rather than ignore, the technologies that are now changing entire industries. Organizations must adapt new ways of working, rather than resist behavior change or tailor the technologies.

JOINING THE BEHAVIOR CHANGE DELTA WITH THE ADOPTION–ADAPTATION STRATEGY MATRIX

Digital-era transformation is an emerging field, of which many authors address a single aspect of the overall challenge.[1] The seminal change management literature is from the enabling era and may not address concerns about the contemporary technology. For example, Kotter has several books that look at the organizational challenges of old. Additionally, in "Who Moved My Cheese," Johnson and Blanchard (1998) looked at individual behavior. Daryl Conner, "Managing at the Speed of Change" (1992), looked at the change from a process perspective. Alvin Toffler's "Future Shock" (1970) looked at the overall environment. More recently and updated for the digital era, a 2017 Forbes Insights and Oracle collaboration attempted to review talent management in the digital era.[2] Still, most of the writing on digital or digital era has been led by consultants at the coalface of rapidly changing and even disappearing industries.[3]

Most change and enabling-era transformation authors have focused on the intersection of business strategy and technological capabilities. They assume that the human issues will somehow be addressed and that technology is subsumed to people and process. Yet "the sheer volume of technologies, processes, and decisions required to build and maintain digital applications and operations means companies can't afford to work in the same old ways."[4]

Modern technology's globally applicable, exogenous processes have replaced the internal procedures workers once used. Technology takers must recognize that technology has ascended to the top of the Behavior Change Delta (comprised of process, people, and technology as the source domains of change in need of change management). The digital era requires a new strategic approach to adopting its distinct technology and adapting to this technology.

The Adoption-Adaptation Strategy Matrix offered four approaches to adopting and adapting to digital-era technology: Maker, Taker, Tinker, and Tailor. The technology taker strategy calls on leaders to adopt digital-era technologies and to adapt their behaviors to optimize these technologies' unique features of interconnectivity and value creation based on information gleaned from the data stream. The Matrix shows that other possible strategies exist for the digital era, but none other than technology taking is ideal for most organizations. Undertaking the other strategies can result in an organization's disintermediation from the marketplace.

These other potential strategic approaches could include tailoring, which involves avoiding all digital-era technology and contently refurbishing and customizing enabling-era relics until these utterly cease to function. Alternatively, with a technology tinkerer, strategy, one could choose to be change adverse and stop the clock – Rip Van Winkle style – for decades as both technology innovation and change management attempts pass by. The truly special might try their hand at technology making, inventing the next Snapchat, Instagram, or FarmVille to revolutionize the virtual world. However, unless your organization's mission is inventing the next farming simulation video game to replace FarmVille, technology taking is the most viable strategy for surviving the dislocations of the digital era.

THE TECHNOLOGY TAKER STRATEGY

The technology taker strategy is threefold: (1) commit to using digital-era technology, even as it increases competitive interconnectivity with external stakeholders (2) match the organization's processes with those required by the technologies used, and (3) require changes in behavior so that management decisions are based on information from data streams. The digital era requires the organization constantly to manage the adoption and adaptation of externally defined, continuously updating technologies. Technology taking and related change management are not a singular event. A strategy as a technology taker requires repeated change management actions that respond to the changes in the technologies themselves. Ultimately, this strategy will put the organization in a better position to create value. The alternative is risking disintermediation as others adopt and adapt across the maker and taker options of the Strategy Matrix.

A technology taker strategy means stating and implementing a preference for using digital-era technologies whenever possible. When an organization accepts the leading technologies, those used by many, the organization does not waste resources trying to customize or tailor these technologies.[5] Efficiency is obtained, and value created, by sharing work processes with others.[6] The organization accepts that, outside its area of expertise, it will not lead on technology creation or be a technology maker.

As a technology taker, it simply will use the technologies that others are using. Nevertheless, the organization and its managers will demonstrate leadership by focusing on delivering their particular mission most effectively at the lowest possible cost and the highest level of value, as demonstrated by goals achieved.

Matching an organization's processes with those provided by the technologies chosen involves the organization's expectation that all its people will use digital-era technologies to implement its policies and conduct its procedures. No one should be immune to becoming a technology taker, and modern tech should be used to its maximum potential. Yet, technology taking is not about the homogenization of businesses or business purposes. Organizations' missions and contributions to the market should remain distinct. However, unless an organization has invented the dominant technology in a market as a technology maker, it cannot sustain a competitive advantage through the in-house specification of the process.

Technology takers should base management decisions on information from the data streams generated by cloud-based tech, as opposed to a leader's guesses or desires. In the digital era, the value is found through the analysis of data streams rather than through process re-engineering. Strategic planning based on data allows leaders to compare their organizations against others in the market to develop unique advantages. Having comparable data and using common processes also permit partnership and interconnectivity among organizations. With better data, and assuming they have the requisite skill, managers can make better decisions, leading to more efficient and less costly business practices.

VIRTUOUS CYCLE OF CHANGE

Leaders should channel the behavior changes digital-era technologies bring and use them to create value for their organizations. A virtuous circle of change can be constructed from five plays. The organization should envision itself as a digital-era organization based on a sustainable business case created at the highest levels of the organization. Having envisioned its technology taking, the organization will need help to be ready for the digital era. Governance involves assessing change preparedness and framing proposed changes within a structured decision-making process, documented in the organization's policies and procedures. Next, leaders must engage with stakeholders about the future state envisioned overall and for specific plans for technology adoption. Then, it is time to invest in and equip the people who will adopt the technologies and change their behaviors. Last, all aspects of technology taking and behavior change (vision, governance, engagement, and other actions) are measured to determine their success at creating value.

Our playbook suggests specifically how technology takers could adapt their organizations to the digital era. These plays contrast with the classical approach to change management. Play 1 offers a new proposition, a change management function, to help organizations engage with and capitalize on digital-era innovations. Play 2 introduces a generally ignored feature, governance, as a basis for guiding and shaping technology taking. Play 3, engage in the digital era, demands the end of reliance on project or singular event mentality. Instead, leaders should issue a clear mandate for technological adoption and adaptation to the digital era, while empowering advocates to echo and amplify the message. Play 4 posits the view that, no matter how intuitive digital-era technologies supposedly are workers will be ill-equipped to use these new applications. Technology adoption must be accompanied by investment in training. Finally, Play 5 argues that organizations must demand real accountability from their managers by measuring technology-taking capabilities to ensure managers are capable and willing to adapt to the digital era. Using these plays will ensure that an organization's uptake of new technology is continuous and behavioral adaptations to this technology create value.

PLAYBOOK LINER NOTES

Play 1 reminded that no one can be an adherent to both the enabling era and the digital era at the same time. The enabling era required a project approach to technology adoption and adaptation: a one-and-done game. The digital era is an iterative game of continual organizational change management.

A permanent change management function (CMF) will be in charge of implementing, but not owning the organization's vision for technology taking and managing the business case for the never-ending cycle of technology adoption and behavioral adaptation. Leadership must own the technology-taking vision and strategy, for which the CMF has two processes: ensuring the adoption of digital-era technologies using a solid and constantly updated business case, and coordinating changes to the organization's culture to accept and use new technology.

A CMF will be made up of elements of a project team and a change management team, permitting it to move beyond the traditional project approach to support change around technology implementation. A chief change officer (CCO) will lead the CMF. As a member of the executive team, the CCO will have the power to promote the technology-taking strategy at her organization. The CCO also can guide the organization to make better decisions about which digital-era technologies to adopt and how specifically to adapt.

Play 2 observed that successful implementation of digital-era technology is predicated on construction of a supportive policy environment and unified governance structure. An organization's mission and strategy for change guarantee that its policies promote, and not encumber, the rapid pace of change and workers' use of digital processes. A review of risks to the organization can help illuminate areas of weakness in organizational policies and where new policies may be required to meet modern or emerging challenges.

A single body should be given ultimate authority over all an organization's policies and procedures. Through its coordination of policies and procedures development and revision, the policy committee assures that the use of new, digital-era systems supports, and does not violate, the organization's mission. The policy committee's leadership over policies and implementing procedures also harmonizes managerial responsibilities for organizational and IT governance.

Adoption of modern technology is an occasion to revise seriously outdated policies and to clarify overly cumbersome procedural requirements. These rationalizations of policies and processes reduce the compliance burden on people and diminish the cost of implementation. However, the organization must anticipate that procedural simplifications may be met with a backlash from managers whose power and expertise was based on control of previous processes.

Play 3 called for engagement and sponsorship to guide an organization through discussions of why the organization is employing a technology taker strategy and how people will be affected by the organization's adoption of digital-era technologies. Sponsors ultimately are responsible for changing the organization's culture from behavior change and technology-resistant to adaptable. To undertake their role, sponsors verbalize the technology-taking strategy. They state a commitment to technology taking; they mandate behavior changes; they order the suspension of outdated processes and require global best practices.

Sponsors themselves model technology taking for their organizations. A digital-era sponsor must be both a technology taker and a modern change manager. Sponsors will use the technologies they are supporting for their organization's adoption. They will change their own behaviors to, for example, make data-driven decisions and to manage information, not people or processes. Technology-taking sponsors will demonstrate in word and deed that no one at their organization may opt out of being a technology taker.

Yet the digital era may exhaust sponsors, as there is no end to change and no end to the need for sponsorship. Sponsors must be strengthened in their work by technology takers at all levels of the organization. To be a technology taker is to be an advocate for technology adoption and adaptation. Advocates advise and motivate sponsors in their work, and advocates also agitate for change within their sphere of influence in the organization.

Play 4 found that organizations can equip people through training to address all areas of the Behavior Change Delta. A training play can improve success in the rise of technology over internal processes and people. Training provides the mental and philosophical preparation necessary to support behavior change.

To be takers of digital-era technologies, users also need training on how to adopt them. Knowledge transfer activities, including training and staff communication, change people's behavior and help inculcate an understanding of new, technology-determined procedures. Training activities and communication efforts explain how and when to use novel processes. Executive management then uses the data generated by the digital-era systems to measure compliance, ensuring the organization's efficiency and effectiveness in meeting its mission.

Training also provides an opportunity for the organization to attract innovative, adaptable employees. If organizations do not provide an opportunity for learning and experimentation with cutting-edge technologies, innovative employees will decide to leave and apply their talents elsewhere. Over time, the staff who are left will be comfortable with the status quo and avoid experimentation. Conversely, technology takers will move to organizations dedicated to constant learning and to encouraging technology taking.

The final *Play 5* examined the question of how one knows if the technology taker strategy is working as intended. What would indicate that an organization was adopting technology and adapting to the requirements of the digital era? The various aspects of the strategy would need to be measured and a determination made of whether the value was being created from the efforts made. Organizations, as led by their change management functions, would need, in particular, to measure managers' own technology adoption and adaptation.

Middle managers are a major obstacle to change.[7] Given the efficiency gains that would come with using digital-era technologies optimally, this opposition to change — over abolishment of a single form or over the elimination of a single wet signature — seems to make little sense. But it is in the minutiae where change resistance makes its last stand. Organizations must measure managers' abilities, alacrity, and success at using the technologies chosen by the organization.

The converse too must be measured: managers' willingness to suspend use of outdated systems and old behaviors. The Behavior Change Delta demonstrates that constantly changing technologies have overtaken the customized processes workers use. The technology taker strategy holds that no one can stand with a foot both in the enabling era and the digital era or in two quadrants of the Adoption–Adaptation Strategy Matrix. Therefore, managers should not try to make their own processes where organization-wide, digital-era ones will suffice. And managers cannot simultaneously make use of old and new processes. A leap to the digital era must occur.

CHANGE LEADERS IN THE DIGITAL ERA

The digital era is a fact of life, and technology taking is the most feasible response. Although other possible strategies exist to respond to the digital era, such as ignoring it completely, adopting its technologies partly, and changing behavior begrudgingly, these plans are likely to result in frustration, low-economic return, or poor achievement of mission. A nonadaptive organization may be replaced in the market by an organization that has kept up with the times. To realize the benefits of digital-era technologies, organizations should not insist that technology should bow to them.

Instead, it is more successful and, frankly, more fun, to be an adopting, adaptive technology taker.

Technology takers hold a dual role as permanent change leaders. The incessant and increasing change of the digital era requires a new form of change leadership. Change projects cannot be singular events with a clear beginning and end built around a specific, desired change. Technology takers must help their organizations and colleagues build the capacity for and acceptance of constant change.

Technology-taking leaders must invest significant effort in envisioning and creating a climate that encourages ways of working via modern technologies, new processes, and with capable people. Through their strategy of always choosing to work with the global-standard technologies, adapting business processes to these technologies, and managing from data, technology takers too will create a new organizational culture. This culture will appreciate the virtues of adopting new ideas and of being flexible responding to modern risks and realities.

The digital era's data streams challenge technology takers to learn a different method of management. Decisions can be based on a comparison of information received against goals set. Organizations can benchmark themselves against other, similar companies using the very same digital-era technologies. Whereas some may feel threatened by the candor data brings to processing times, policy compliance, worker efficiency, and the like, technology takers thrive on transparency.

Change leaders apply this enthusiasm for openness and accountability to all aspects of their work. They encourage from all levels of the organization ideas about technologies to adopt and behaviors to adapt. Although business process experts may not be the most highly placed in the organization, BPEs are the super-users of digital-era technology and have an important perspective on what works and what does not. Technology takers would be wise to seek the opinion of the BPEs on how to reduce the distance between ideas and implementation deeds, and between people and processes.[8] The BPEs can indicate to an organization's leaders what new technologies are dominating the marketplace and how these could contribute to the organization. As adept leaders of the digital era, technology takers then should study these technologies, try them out, and apply them to their organizations.

Technology takers are thus reliant on technology and people to generate ideas, they look to their organizations' missions to direct them, and they rely on their policies to reign in their impulses. Provided they have followed the guiding principles of the technology taking strategy, what technology takers do not do is define the processes by which their organizations work. To be a technology taker is to be agnostic to the processes chosen to implement policy, to achieve the mission, and to generate value. Users of digital-era technologies trust that these technologies' processes truly are the world's best and that organizational time trying to recreate or customize them would be wasted effort.

As much as technology takers are process agnostics, they seek self-enlightenment from their inherent biases and mistaken beliefs. Are you really working effectively? Is your work efficient? Is it ethical? Or are you wasting time and effort in your use of new technologies? Digital-era technologies, with their data streams and interactivity, can help answer these questions.

When confronted with the data, Apple's CEO Tim Cook was surprised to see the amount of time he spent each day and week on apps on his iPhone.[9] And he realized that he was too tied to his Apple devices; they were controlling him and not the other way around.[10] Enlightened by information, Cook now plans to use new Apple product features to change his behavior and to limit the time he wastes on them — so that he can spend his time creating more value for Apple.

But data from digital-era data streams can only inform human decisions; technology has not taken over humans' decision-making just yet. An entire industry has sprung up to provide artificial intelligence-enabled applications to remove biases from hiring.[11] These companies claim to use AI to scan resumes or surveys to identify a candidate's true strengths or weaknesses and not to discriminate against candidates based on immutable characteristics. However, a technology is a product of its maker and can reflect and reinforce the biases and prejudices of the humans who made them.[12] The AI applications for hiring are not really neutral and can increasingly optimize for bias.[13] If the algorithm learns what a "good" hire looks like based on biased data, it will make biased hiring decisions.[14] Those who do not understand the technologies they are using may be duped into believing that the technology can correct for human inabilities and frailties; in fact, technologies merely mirror what we already are.

No managerial decision can be based solely on an algorithm. Data generated from any digital-era technology can add evidence to a human being's decision, but the decision still must be human-made. Technology takers can be guided by the plays in this book to be ready, engaged, educated, and supported in the decisions they must make in the digital era.

THE TECHNOLOGY TAKERS

Being a technology taker requires considerable effort and is somewhat fraught with risk. Should a leader develop another priority than implementing the technology taking strategy, the technology taker may appear as a wild-eyed prophet, clamoring for fidelity to an approach no one else takes seriously. If, due to no fault of the technology taker's, an organization should fail in its adoption of a technology or its ability to change, the technology taker may be targeted for blame or serve as the lightning rod for the organization's frustration.

Given the drawbacks to technology taking, why bother? Perhaps it would be better to decide to be a technology tinkerer and to let the digital era gently pass you by. As a final argument for technology taking, we suggest that this strategy will provide the most options for choice within the market and the greatest opportunity for fun.

An entire generation of American children grew up skipping rope to the McDonald's Big Mac Song: "two all-beef patties, special sauce, lettuce, cheese, pickles, onions, on a sesame seed bun."[15] While serving as a prototype example of effective marketing, the Big Mac Song also set out exactly what was on a Big Mac. A Big Mac was not made to order; it came with its special sauce and its lettuce, cheese, and pickles. Theoretically, it was possible to ask for no onions; but, the sandwich delivered to you might have onions anyway. You had been warned. There were no requesting toppings of avocados, feta cheese, or heirloom tomatoes; no one ate these in the 1970s and 1980s; and, more to the point, they were not part of the Big Mac ingredients list that everyone knew.

Enter the digital era, and McDonald's now has touchscreen ordering devices in its restaurants and an app for ordering on mobile devices. These technologies permit the customer to choose from a plethora of ways to craft her Big Mac.[16] One could request pico de gallo and guacamole. Or bacon, garlic, habanero ranch sauce, or white cheddar (instead of generic "cheese"). The combinations are endless. Once the customer has placed her order for a unique and special hamburger, McDonald's partnership with UberEats will deliver that Big Mac straight to the customer's home.

The technology taker is no longer roped to a sing-song definition of a Big Mac nor to shoving paper from one side of her desk to the other. Using digital-era technologies, the taker can free herself from cumbersome, time-consuming manual processes and open up a world of choice. Time can be spent contemplating the many novel ways of communicating and interacting with people around the world or in sorting through data to see what they are saying.

Technology taking also is fun. Digital-era technologies make learning to use them an enjoyable and engaging process. Mastering how really to use a new technology will distinguish you from those who are still unaware of that technology's existence. It is a buzzy, in-the-know feeling to have a little bit more knowledge than those around you. You will understand what is being talked about on the nightly news and what forces are moving markets. Your skills will soon be in demand as the industry standard, and the rest will be struggling to catch up. You will be able to seek out a work environment where your technology-taking philosophy and abilities are valued and appreciated.

As they adopt a new language, habits, and philosophy, technology takers will start to form their own community. Perhaps you will be the first technology taker at your organization. But you will soon find a second like-minded individual, and then, a third. You will have in common your interest in modern technologies and how to use them, and you will also share an appreciation for the challenges of being an advocate for change in an often-hostile environment. When the going gets particularly tough, your fellow technology takers will encourage one another to follow the playbook and stick with the plan to achieve long-term change and success.

Interaction with the merry band of technology takers generates anticipation of change and of the possibilities of the digital era; they are the psychological equivalent of the new iPhone smell. They build a sense of hope, rather than resentment, of technologically driven changes to come. Leaders start and advocates join in the chant: we are now operating as digital-era, technology takers.

At its core, though, the attraction of technology taking is almost primal. Technology taking delights in using the latest, greatest gadget or app with all its bells and whistles. Who doesn't want to be cutting-edge? You will glimpse the future before others do, and you will already be comfortable with tomorrow's changes by the time others catch up to you. By being a technology taker, you will experience the magic of what the future soon will bring to us all.

NOTES

1. Morakanyane, R., Grace, A., & Oreilly, P. (2017). Conceptualizing digital transformation in business organizations: A systematic review of literature. *Digital Transformation – From Connecting Things to Transforming Our Lives*. doi:10.18690/978-961-286-043-1.30

2. Desmet, D., Löffler, M., & Weinberg, A. (September, 2016). *Modernizing IT for a digital-era*. Retrieved from https://www.mckinsey.com/business-functions/digital-mckinsey/our-insights/modernizing-it-for-a-digital-era

3. Ibid.

4. Ibid.

5. "Self-adaptive or autonomic systems promise to move this complexity [caused from customization of systems] from humans into the software itself, thus reducing software maintenance cost, improving performance of systems, customer satisfaction, and so on": Lapouchnian, A. (2011). *Exploiting requirements variability for software customization and adaptation*. Retrieved from https://www.semanticscholar.org/paper/Exploiting-Requirements-Variability-for-Software-Lapouchnian-Easterbrook/ae7bc1b0f4a959ec4337771ee2532cf04e439994

6. "Relational-specific resource commitment [or, cooperation on work processes] creates a mutually beneficial environment that allows the firm to realize the potential value attainable through collaboration": Wu, F., & Cavusgil, S. T. (2006). Organizational learning, commitment, and joint

value creation in interfirm relationships. *Journal of Business Research,* *59*(1), 81–89. doi:10.1016/j.jbusres.2005.03.005

7. Noria Corporation. (August 09, 2007). *Middle managers are biggest obstacle to lean enterprise.* Retrieved from https://www.reliableplant.com/ Read/7751/middle-managers-lean

8. Ostrom, V. (1997) *The meaning of democracy and the vulnerabilities of democracies: A response to tocqueville's challenge.*

9. Cook, T. (June 4, 2018). *Tim Cook reveals his tech habits: I use my phone too much.* Retrieved from http://money.cnn.com/2018/06/04/technology/apple-tim-cook-screen-time/index.html

10. Ibid.

11. Wachter-Boettcher, S. (October 25, 2017). *AI recruiting tools do not eliminate bia*s. Retrieved from http://time.com/4993431/ai-recruiting-tools-do-not-eliminate-bias/

12. O'Neil, G. M. (February 22, 2017). *Hiring algorithms are not neutral.* Retrieved from https://hbr.org/2016/12/hiring-algorithms-are-not-neutral

13. Wachter-Boettcher, S. (October 25, 2017). *AI recruiting tools do not eliminate bias.* Retrieved from http://time.com/4993431/ai-recruiting-tools-do-not-eliminate-bias/

14. O'Neil, G. M. (February 22, 2017). *Hiring algorithms are not neutral.* Retrieved from https://hbr.org/2016/12/hiring-algorithms-are-not-neutral

15. Clifford, S. (July 17, 2008). *Remember '2 All-Beef Patties?' McDonald's hopes you do.* Retrieved from https://www.nytimes.com/ 2008/07/17/business/media/17adco.html

16. *McDonald's Burgers: Hamburgers & cheeseburgers* | McDonald's. (n.d.). Retrieved from https://www.mcdonalds.com/us/en-us/full-menu/burgers.html

BIBLIOGRAPHY

ACHE Healthcare Executive. (2018). Competencies assessment tool. *Research & Resources*. Retrieved from https://www.ache.org/newclub/resource/competencies.cfm

Accenture. (n.d). *Creating the best customer experience*. Retrieved from https://www.accenture.com/us-en/interactive-index

Ahadi, H. (2004). An examination of the role of organizational enablers in business process reengineering and the impact of information technology. *Information Resources Management Journal, 17*(4), 1–19. doi:10.4018/irmj.2004100101

Al-Badi, A., Tarhini, A., & Al-Kaaf, W. (2017). Financial incentives for adopting cloud computing in higher educational institutions. *Asian Social Science, 13*(4), 162. doi:10.5539/assv13n4p162

Alexandre, A. (2018, April 12). Belgium contributes to world food programme blockchain project. *Cointelegraph*, Retrieved from https://cointelegraph.com/news/belgium-contributes-to-world-food-programme-blockchain-project

Alsher, P. (2013, October 29). The CAST of characters for implementing organizational changes. *IMA's implementing organizational changes at speed blog*. Retrieved from https://www.imaworldwide.com/blog/bid/189157/The-CAST-of-Characters-for-Implementing-Organizational-Changes

Alsher, P. (2013, October 29). *The CAST of characters for implementing organizational changes*. Retrieved from https://www.imaworldwide.com/blog/bid/189157/The-CAST-of-Characters-for-Implementing-Organizational-Changes

American College of Health Executives. (2018). ACHE healthcare executive 2018 competencies assessment tool. *American College of Health*

Executives. Retrieved from https://www.ache.org/newclub/resource/
competencies.cfm

Anderson, H. C. (1949). *The emperor's new clothes*. Boston: Houghton
Mifflin Co.

Andriole, S. (2015, October 08). What C-suite executives need to know
about digital strategy and emerging technologies. *Forbes*, Retrieved from
https://www.forbes.com/sites/steveandriole/2015/10/07/analytics-iot-social-
location-security-how-to-all-get-along-a-note-to-c-suiters/#dfbc646ab27a

Andriole, S. (2018, April 13). Implement first, ask questions later (or not
at all). *MITSloan Management Review*. Retrieved from https://
sloanreview-mit-edu.cdn.ampproject.org/c/s/sloanreview.mit.edu/article/
implement-first-ask-questions-later-or-not-at-all/amp

Andriole, S. (2015, October 08). *What C-Suite executives need to know
about digital strategy and emerging technologies*. Forbes.com. Retrieved from
https://www.forbes.com/sites/steveandriole/2015/10/07/analytics-iot-social-
location-security-how-to-all-get-along-a-note-to-c-suiters/#dfbc646ab27a

Anthony, S. (2014, July 23). First mover or fast follower? *Harvard
Business Review*, Retrieved from https://hbr.org/2012/06/first-mover-or-
fast-follower

Appian. (n.d). *Top 10 digital transformation trends for 2018*. Retrieved
from https://sf.tradepub.com/free/w_appf228/

Appian. (2018). Top 10 digital transformation trends for 2018.
TradePub.com. Retrieved from https://sf.tradepub.com/free/w_appf228/

Apple. (2017, December 28). A message to our customers. *Apple*.
Retrieved from https://www.apple.com/iphone-battery-and-performance/

Apple. (2017, December 28). *A message to our customers*. Retrieved from
https://www.apple.com/iphone-battery-and-performance/

Armenakis, A. A., Harris, S. G., & Mossholder, K. W. (1993). Creating
readiness for organizational change. *Human Relations*, 46(6), 681–703.
doi:10.1177/001872679304600601

AT&T Profile. (n.d). *AT&T Company Website*. Retrieved from
http://about.att.com/sites/company_profile

Avila, O., & Garcés, K. (2016). Change management support to preserve business—information technology alignment. *Journal of Computer Information Systems, 57*(3), 218–228. doi:10.1080/08874417.2016.1184006

Barreau, D. (2001). The hidden costs of implementing and maintaining information systems. *The Bottom Line, 14*(4), 207–213. doi:10.1108/08880450110408481

Beattie, A. (2018, May 13). Data protectionism: The growing menace to global business. *Financial Times.* Retrieved from https://www.ft.com/content/6f0f41e4-47de-11e8-8ee8-cae73aab7ccb

Beckley, A. M. (2015, August 07). How the Cloud is Changing the Role of Technology Leaders. *Wired.* Retrieved from https://www.wired.com/insights/2013/09/how-the-cloud-is-changing-the-role-of-technology-leaders/

Benko, C., & Donovan, J. (2016, October 07). Inside AT&T's radical talent overhaul. *Harvard Business Review.* Retrieved from https://hbr.org/2016/10/atts-talent-overhaul

Bezos, J. (1997). Letter to Shareholders. *U.S. securities and exchange commission archives.* Retrieved from https://www.sec.gov/Archives/edgar/data/1018724/000119312518121161/d456916dex991.htm

Birkinshaw, J., & Heywood, S. (n.d.). Putting organizational complexity in its place. *McKinsey Insights,* Retrieved from https://www.mckinsey.com/business-functions/organization/our-insights/putting-organizational-complexity-in-its-place

Blake, M. (2018, April 13). On disability, Twitter is better late than never. *CNN.* Retrieved from https://www.cnn.com/2018/04/13/opinions/twitter-changes-terms-on-disability-blake-opinion/index.html

Blons, E. (2018, February 17). Change Management in the Era of Artificial Intelligence: Building trust through transparency and accountability. Linkedin Pulse. Retrieved from https://www.linkedin.com/pulse/change-management-era-artificial-intelligence-building-blons/

Bolles, R. N. (1981). *Three boxes of life and how to get out of them.* California: Ten Speed Press.

Bonnet, D. (2014, November 14). Convincing employees to use new technology. *Harvard Business Review*, Retrieved from https://hbr.org/2014/09/convincing-employees-to-use-new-technology

Boston Consulting Group. (n.d). Digital Transformation – Strategy for Digitizing the Business. *Boston Consulting Group*. Retrieved from https://www.bcg.com/capabilities/technology-digital/digital.aspx

Bridges, W. (2017). *Managing transitions: Making the most of change.* Boston: Da Capo Lifelong Books.

Brothers, C. (2005). *Language and the pursuit of happiness: A new foundation for designing your life, your relationships & your results.* Naples, FL: New Possibilities Press.

Brown, S. A. (n.d). A History of the Bar Code. *Economic History Services.* Retrieved from https://eh.net/encyclopedia/a-history-of-the-bar-code/

Calaway, L. (2017, May 23). Sink the Boats and Burn the Bridges! *Expressworks*. Retrieved from http://www.expressworks.com/sharepoint-adoption/sink-the-boats-and-burn-the-bridges

Caminiti, S. (2018, March 13). AT&T's $1 billion gambit: Retraining nearly half its workforce for jobs of the future. *CNBC*. Retrieved from https://www.cnbc.com/2018/03/13/atts-1-billion-gambit-retraining-nearly-half-its-workforce.html

Campbell, A., Goold, M., & Alexander, M. (1995). Corporate strategy: The quest for parenting advantage. *Harvard Business Review*, 73(2), 120–132.

Cannon, W. B. (1929). *Bodily changes in pain, hunger, fear and rage: An account of recent researches into the function of emotional excitement.* New York: D. Appleton.

Cava, M. D., & Jones, C. (2016, April 22). For older CEOs, issue is knowing when to bow out. *USA Today*. Retrieved from https://www.usatoday.com/story/money/business/2016/04/19/older-ceos-issue-knowing-when-bow-out/83114728/

CC Group. (n.d). Agriculture Blockchain Technology. (n.d.). *CC Group*. Retrieved from https://ccgrouppr.com/practical-applications-of-blockchain-technology/sectors/agriculture/

Charlotte Center City. (n.d). Center City 2020 Vision Plan. *Charlotte Center City*. Retrieved from https://www.charlottecentercity.org/center-city-initiatives-2/plans/2020-vision-plan/

Charlotte, North Carolina Population 2018. (2018, June 12). *World Population Review*. Retrieved from http://worldpopulationreview.com/us-cities/charlotte-population/

Chatterjee, S. (2018, May 14). HSBC says performs first trade finance deal using single blockchain. *Reuters UK*. Retrieved from https://uk.reuters.com/article/uk-hsbc-blockchain/hsbc-says-performs-first-trade-finance-transaction-using-blockchain-idUKKCN1IF03H

CHG. (2009, December 22). *Healthcare services placed among global elite corporate training programs*. New-Medical.net. CHG Healthcare Services placed among global elite corporate training programs. (2009, December 22). *New Medical*. Retrieved from https://www.news-medical.net/news/20091222/CHG-Healthcare-Services-placed-among-global-elite-corporate-training-programs.aspx

Christensen, J. (2011, August). The Decision to Internally Generate or Outsource Risk Management Activities. August 2011. *Bond University*. Retrieved from https://epublications.bond.edu.au/cgi/viewcontent.cgi?article=1095&context=theses

Clarke, A. (2017). Digital government units: Origins, orthodoxy and critical considerations for public management theory and practice. *SSRN Electronic Journal*. doi:10.2139/ssrn.3001188

Clifford, S. (2008, July 17). Remember '2 All-Beef Patties?' McDonald's hopes you do. *The New York Times*. Retrieved from https://www.nytimes.com/2008/07/17/business/media/17adco.html

CNN Video. (2018, April 11). Confusing questions Congress asked Zuckerberg. *CNN*. Retrieved from https://www.cnn.com/videos/cnnmoney/2018/04/11/facebook-zuckerberg-confusing-questions-congress-cnnmoney-orig.cnnmoney

Cohan, P. (2018, February 15). Apple: Warren Buffett's Second Big Bad Tech Bet. *Forbes*. Retrieved from https://www.forbes.com/sites/petercohan/2018/02/15/apple-warren-buffetts-second-big-bad-tech-bet/#77b054e667ff

Conner, D. (2006). *Managing at the speed of change: How resilient managers succeed and prosper where others fail*. New York: Random House.

Cook, T. (2018, June 4). Tim Cook reveals his tech habits: I use my phone too much. *CNN Money*. Retrieved from http://money.cnn.com/2018/06/04/technology/apple-tim-cook-screen-time/index.html

Cosgrove, E. (2017, August 24). Cargill Invests in Predictive Ag 'Data Refinery' Descartes Labs' $30m Series B. *AGfunders News*. Retrieved from https://agfundernews.com/descartes-raise.html

Cotula, L. (2013). The new enclosures? Polanyi, international investment law and the global land rush. *Third World Quarterly, 34*(9), 1605–1629. doi:10.1080/01436597.2013.843847

Cox, T. (2008, October 21). The kitty site that's a phenomenon. *The Times*. Retrieved from https://www.thetimes.co.uk/article/the-kitty-site-thats-a-phenomenon-5nh9bfpskzg

Creasey, T. (n.d.). The costs & risks of poorly managed change. *Prosci Change Management Blog*. Retrieved from http://blog.prosci.com/the-costs-risks-of-poorly-managed-change

Creasey, T. (n.d.). Latest data and key considerations for the CMO. *Prosci Change Management Blog*. Retrieved from http://blog.prosci.com/Latest-Data-and-Key-Considerations-for-the-CMO

Davies, H. (2015, December 11). Ted Cruz campaign using firm that harvested data on millions of unwitting Facebook users. *The Guardian*. Retrieved from https://www.theguardian.com/us-news/2015/dec/11/senator-ted-cruz-president-campaign-facebook-user-data

Deloitte Development LLC. (2013). Boards and management teams – Illustrative governance operating model. *Deloitte*. Retrieved from https://www2.deloitte.com/content/dam/Deloitte/global/Documents/Financial-Services/dttl-fsi-US-FSI-Developinganeffectivegovernance-031913.pdf

Desmet, D., Löffler, M., & Weinberg, A. (2016, September). Modernizing IT for a digital-era. *McKinsey Insights*. Retrieved from https://www.mckinsey.com/business-functions/digital-mckinsey/our-insights/modernizing-it-for-a-digital-era?cid=eml-web

Dewnarain, G., O'Connell, D., & Gotta, M. (2017, October 6). SWOT: Slack, Worldwide. *Gartner*. Retrieved from https://www.gartner.com/doc/reprints?id=1-4K5I75F&ct=171108&st=sb

Dickens, C. (1859). *A tale of two cities*. London: Chapman and Hall.

Draznin, H. (2017, June 30). Jimmy Choo Co-Founder: "Society is better off when women earn equal". *CNN Money*. Retrieved from http://money.cnn.com/2017/06/30/smallbusiness/tamara-mellon-jimmy-choo/index.html

Dunleavy, P., Bastow, S., Margetts, H., & Tinkler, J. (2006). *Digital era governance*. Oxford: Oxford University Press.

Dunn, L. E. (2017, June 05). Women in business Q&A: Tamara mellon. *Huffington Post*, Retrieved from https://www.huffingtonpost.com/entry/women-in-business-qa-tamara-mellon_us_59357964e4b0f33414194bf4

Eadicicco, L. et al. (2017, April 03). TIME's 20 most successful technology failures of all time. *Time*. Retrieved from http://time.com/4704250/most-successful-technology-tech-failures-gadgets-flops-bombs-fails/

Eccles, R. G. (1991). The performance measurement Manifesto. *Harvard Business Review*, 69(1), 131–137.

Elton, J., & Roe, J. (1998). Bringing discipline to project management. *Harvard Business Review*, 76(2), 153–160.

Eshet-Alkalai, Y. (n.d.). A holistic model of thinking skills in the digital era. *Encyclopedia of distance learning*, Second Edition, 1088-1093. https://doi.org/10.4018/978-1-60566-198-8.ch154

Everett, C. (2017, January 19). How culture change has to underpin success in digital transformation. *Computer Weekly*. Retrieved from https://www.computerweekly.com/feature/How-culture-change-has-to-underpin-success-in-digital-transformation

Ewenstein, B., Smith, W., & Sologar, A. (2015, July). Changing change management. *McKinsey Insights*. Retrieved from https://www.mckinsey.com/featured-insights/leadership/changing-change-management

Facebook and the meaning of share ownership. (2017, September 30). *Economist*. Retrieved from https://www.economist.com/business/2017/09/30/facebook-and-the-meaning-of-share-ownership

Farnel, F. (1994). *Lobbying: Strategies and techniques of intervention.* Paris: Editions d'Organisation.

Fisher, M. (1991). *The millionaire's book of quotations.* London: Thorsons.

Flanding, J., & Grabman, G. (2016). Change management in the cloud: The case for digital era governance. Presented at Cutter Summit: Unlock Digital Transformation, Cambridge, MA.

Foss, N. J., & Lindenberg, S. (2013). Microfoundations for strategy: A goal-framing perspective on the drivers of value creation. *Academy of Management Perspectives, 27*(2), 85–102. doi:10.5465/amp.2012.0103

Fowler, G. A. (2018, April 05). What if we paid for Facebook - Instead of letting it spy on us for free? *Washington Post.* Retrieved from https://www.washingtonpost.com/news/the-switch/wp/2018/04/05/what-if-we-paid-for-facebook-instead-of-letting-it-spy-on-us-for-free/

Freifeld, L. (2013, October 22). *L&D best practices: Technology and technical development.* Retrieved from https://trainingmag.com/content/ld-best-practices-technology-and-technical-development

Fruhlinger, J., & Wailgum, T. (2017, July 10). 15 Famous ERP disasters, dustups and disappointments. *CIO.* Retrieved from https://www.cio.com/article/2429865/enterprise-resource-planning/enterprise-resource-planning-10-famous-erp-disasters-dustups-and-disappointments.html

Fuller, L., & Smith, V. (1991). Consumers reports: Management by customers in a changing economy. *Work, Employment and Society, 5*(1), 1–16. doi:10.1177/0950017091005001002

Galang, J. (2016, July 14). *Calgary-born Robert Opp leading UN World Food Programme's new innovation division.* Retrieved from https://betakit.com/calgary-born-robert-opp-leading-un-world-food-programmes-new-innovation-division/

Regulation (EU). (2016). 2016/679, of the European Parliament and the Council of 27 April 2016 on the protection of natural persons with regard to the processing of personal data and on the free movement of such data, and repealing Directive 95/46/EC (General Data Protection Regulation), O.J. (L 119) 1 (hereafter "GDPR").

Gavett, G. (2016, August 04). How self-service kiosks are changing customer behavior. *Harvard Business Review*. Retrieved from https://hbr.org/2015/03/how-self-service-kiosks-are-changing-customer-behavior

Glon, R. (2017, October 22). *How does Uber work? Here's how the app lets you ride, drive, or both.* DigitalTrends.com. Retrieved from https://www.digitaltrends.com/cars/how-does-uber-work/

Gofore. (2017, December 18). *Without risk there is no innovation.* Retrieved from https://gofore.com/en/without-risk-no-innovation/

Goodall, M. B., Buckingham, M., & Ashkenas, R. (2015, November 16). Reinventing performance management. *Harvard Business Review*, Retrieved from https://hbr.org/2015/04/reinventing-performance-management

Goran, J., LaBerge, L., & Srinivasan, R. (n.d.). Culture for a digital age. *McKinsey Insights*, Retrieved from https://www.mckinsey.com/business-functions/digital-mckinsey/our-insights/culture-for-a-digital-age

Graham-Harrison, E., & Cadwalladr, C. (2018, March 17). Revealed: 50 million Facebook profiles harvested for Cambridge Analytica in major data breach. *The Guardian*. Retrieved from https://www.theguardian.com/news/2018/mar/17/cambridge-analytica-facebook-influence-us-election

Greene, J. A., & Kesselheim, A. S. (2010). Pharmaceutical marketing and the new social media. *New England Journal of Medicine, 363*(22), 2087–2089. doi:10.1056/nejmp1004986

Grzinich, J. C., Thompson, J. H., & Sentovich, M. F. (1997) *Implementation of an integrated product development process for systems,* https://doi.org/10.1109/PICMET.1997.653448

Guidestar.org. (n.d.) *African Mothers health initiative.* Retrieved from https://www.guidestar.org/profile/26-0423197

Günther, W. A., Mehrizi, M. H., Huysman, M., & Feldberg, F. (2017). Debating big data: A literature review on realizing value from big data. *The Journal of Strategic Information Systems, 26*(3), 191–209. doi:10.1016/j.jsis.2017.07.003

Haas, J. B. (2016, May 10). Increase your return on failure. *Harvard Business Review*, Retrieved from https://hbr. org/2016/05/increase-your-return-on-failure

Hall, B. H., & Khan, B. (2002, November). *Adoption of new technology.* Retrieved from https://eml.berkeley.edu/~bhhall/papers/HallKhan03% 20diffusion.pdf

Haendly, M. (2016, April 26). *5 tangible benefits of digital transformation.* Retrieved from https://sapinsider.wispubs.com/Assets/ Articles/2016/April/SPI-5-Tangible-Benefits-of-Digital-Transformation

Hammer, M., & Champy, J. (2009). *Re-engineering the corporation.* New York: HarperCollins Publishers..

Hankin, A. (2018, June 28). *9 companies Amazon is killing.* Retrieved from https://www.investopedia.com/news/5-companies- amazon-killing/

Hannam, K. (2017, December 13). *GE wants to know why former CEO Jeff Immelt traveled the world with a spare jet.* Fortune.com. Retrieved from http://fortune.com/2017/12/13/ge-investigation-jeff-immelt-spare-jet/

Harnessing the Energy of Change Champions. (n.d.). The Clemmer Group. Retrieved from https://www.clemmergroup.com/articles/ harnessing-energy-change-champions

Harvard Business Review Staff. (2016, November 03). The four phases of project management. *The Harvard Review.* Retrieved from https://hbr.org/ 2016/11/the-four-phases-of-project-management

Heath, A. (2017, June 22). Facebook has a new mission statement: 'To bring the world closer together.' *Business Insider.* Retrieved from http://www.businessinsider.com/new-facebook-mission-statement- 2017-6

Heng, Y., & Aljunied, S. M. A. (2015). Can small states be more than price takers in global governance? *Global Governance, 21*(3), 435.

Herger, M. (2006, June 4). *What is a business process expert, really?* Retrieved from https://blogs.sap.com/2006/06/04/what-is-a-business- process-expert-really/

Hern, A. (2014, November 19). Uber investigates top executive after journalist's privacy was breached. *The Guardian.* Retrieved from https:// www.theguardian.com/technology/2014/nov/19/uber-investigates-top- executive-after-journalists-privacy-was-breached

Hirt, M., & Willmott, P. (2014, May). Strategic principles for competing in the digital era. *McKinsey Insights*. Retrieved from https://www.mckinsey.com/business-functions/strategy-and-corporate-finance/our-insights/strategic-principles-for-competing-in-the-digital-age

Hopkins, J. (2017, July 10). *Millennial managers: A guide for successful management.* Forbes.com. Retrieved from https://www.forbes.com/sites/jamiehopkins/2017/07/08/millennials-managers-a-guide-for-successful-management/#34ec6a6e2ac3

Hsu, T., & Kang, C. (2018, March 26). Demands grow for Facebook to explain its privacy policies. *NY Times*, Retrieved from https://www.nytimes.com/2018/03/26/technology/ftc-facebook-investigation-cambridge-analytica.html

Hunt, E., & Banaji, M. R. (1988). The Whorfian hypothesis revisited: A cognitive science view of linguistic and cultural effects on thought. *Indigenous Cognition: Functioning in Cultural Context, 41*, 57–84. doi:10.1007/978-94-009-2778-0_5

Institute of Risk Management. (n.d.). *Our story.* Retrieved from https://www.theirm.org/about/our-story.aspx

International Organization for Standardization. (2009, November). *Risk management – Principles and guidelines.* Retrieved from https://www.iso.org/standard/43170.html

Irrera, A. (2017, April 10). Banks scramble to fix old systems as IT 'cowboys' ride into sunset. *Reuters*. Retrieved from https://www.reuters.com/article/us-usa-banks-cobol/banks-scramble-to-fix-old-systems-as-it-cowboys-ride-into-sunset-idUSKBN17C0D8

Jackson, H. L. (2015, July 13). The hard side of change management. *Harvard Business Review*. Retrieved from https://hbr.org/2005/10/the-hard-side-of-change-management

Jain, A., & Beale, A. (2017). Developing a business case for digital investments in health and social care. *International Journal of Integrated Care, 17*(5), 316. doi:10.5334/ijic.3633

Järild, A. (n.d.). *How digital disintermediation is disrupting food and financial advice.* Retrieved from https://blog.

thinque.com.au/how-digital-disintermediation-is-disrupting-food-and-financial-advice

Javan, J. (2018). Share knowledge for learning, not marketing. *UN Staff System College Blog*, Retrieved from http://www.unssc.org/news-and-insights/blog/share-knowledge-learning-not-marketing/

Johansson, A. (2015, August 20). *FOMO marketing in the age of social media*. Retrieved from https://www.relevance.com/fomo-marketing-in-the-age-of-social-media/

Kantor, J., & Streitfeld, D. (2015, August 15). Inside Amazon: Wrestling big ideas in a bruising workplace. *NY Times*, Retrieved from https://www.nytimes.com/2015/08/16/technology/inside-amazon-wrestling-big-ideas-in-a-bruising-workplace.html

Kaufman, L. (2014, February 1). Chasing their star, on YouTube. *NY Times*. Retrieved from https://www.nytimes.com/2014/02/02/business/chasing-their-star-on-youtube.html

Keller, S., & Aiken, C. *The inconvenient truth about change management*. New York City: McKinsey & Company. Retrieved at: http://www.aascu.org/corporatepartnership/McKinseyReport2.pdf

Kenny, G. (2018, April 30). Your strategic plans probably aren't strategic, or even plans. *Harvard Business Review*, Retrieved from https://hbr.org/2018/04/your-strategic-plans-probably-arent-strategic-or-even-plans

Kim, K., Jung, S., Hwang, J., & Hong, A. (2017). A dynamic framework for analyzing technology standardisation using network analysis and game theory. *Technology Analysis & Strategic Management*, 30(5), 540–555. doi:10.1080/09537325.2017.1340639

Knol, W. H. C., & Stroeken, J. H. M. (2001). The diffusion and adaption of information technology in small and medium sized enterprises through IT scenarios. *Technology Analysis & Strategic Management*, 13(2), 227–246. doi:10.1080/09537320123815

Knowles, M. S., F, H. I., & Swanson, R. A. (2015). *The adult learner: The definitive classic in adult education and human resource development*. Milton Park, Abingdon, Oxon: Routledge.

Kolbjornsrud, V., Amico, R., & Thomas, R. J. (2016, November 2). How artificial intelligence will redefine management. *Harvard Business Review*, Retrieved from https://hbr.org/2016/11/how-artificial-intelligence-will-redefine-management

Kopp, R. (n.d.). *Defining Nemawashi*. Retrieved from http://www.japanintercultural.com/en/news/default.aspx?newsID=234

Kotter, J. P. (1996) *Leading change*. Boston: Harvard Business School Press.

Kotter, J. P. (1995). Leading change: Why transformation efforts fail. *Harvard Business Review*, 73(2), 59–67.

Kotter, J. P. (2007, January). Leading change: Why transformation efforts fail. *Harvard Business Review*, Retrieved from https://hbr.org/2007/01/leading-change-why-transformation-efforts-fail

Kouzes, J. M., & Posner, B. (2009, January). To lead, create a shared vision. *Harvard Business Review*, Retrieved from https://hbr.org/2009/01/to-lead-create-a-shared-vision

Kroc, R., & Anderson, R. (1987). *Grinding it out: The making of McDonald's*. London: St. Martin's Griffin..

Kupersmith, K., Mulvey, P., & McGoey, K. (n.d.). *How to write a cost/benefit analysis for a business case*. Retrieved from https://www.dummies.com/business/business-strategy/how-to-write-a-costbenefit-analysis-for-a-business-case/

Lally, P. et al. (2009). How are habits formed: Modelling habit formation in the real world. *European Journal of Social Psychology*, 40(6), 998–1009. doi:10.1002/ejsp.674

Lambert, F. (2018, May 31). *Tesla Model 3 stopping distance improvements confirmed in new test, Musk says UI/ride comfort improvements coming*. Retrieved from https://electrek.co/2018/05/30/tesla-model-3-stopping-distance-improvements-new-test-ui-ride-comfort-road-noise/

Lapouchnian, A. (2011, June 01). *Exploiting requirements variability for software customization and adaptation*. Retrieved from https://tspace.library.utoronto.ca/handle/1807/27586

Lattuch, F., & Seifert, A. (2014). Insights from change management consulting: Linking the hard and soft side of change with heuristics. In *Management of Permanent Change (pp. 177–194)*. New York: Springer.

Lawrynuik, S. (2018, January 03). Albertan born on a grain farm to rethink how World Food Programme's humanitarian aid is delivered. *CBC News*, Retrieved from http://www.cbc.ca/news/canada/calgary/robert-opp-world-food-progamme-innovation-alberta-1.4471461

Lawson, E., & Price, C. (2003, June). The psychology of change management. *McKinsey Insights*, Retrieved from https://www.mckinsey.com/business-functions/organization/our-insights/the-psychology-of-change-management

Leary, K. (2018, February 27). *The verdict is in: AI outperforms human lawyers in reviewing legal documents*. Futurism.com. Retrieved from https://futurism.com/ai-contracts-lawyers-lawgeex/

Lessick, S., & Kraft, M. (2017). Facing reality: The growth of virtual reality and health sciences libraries. *Journal of The Medical Library Association*, *105*(4), 407–417. doi:10.5195/jmla.2017.329

Levin, M. (2017, March 30). *Why great leaders (Like Richard Branson) inspire instead of motivate*. Inc.com. Retrieved from https://www.inc.com/marissa-levin/why-great-leaders-like-richard-branson-inspire-instead-of-motivate.htm

Levy, A. (2018, April 27). *Amazon's sellers are going global, helping the company generate big profits*. CNBC.com. Retrieved from https://www.cnbc.com/2018/04/26/amazon-25-percent-of-third-party-sales-came-from-global-sellers.html

Lewin, K. (1935) *A dynamic theory of personality*. New York: McGraw-Hill.

Lewin, K. (1936) *Principles of topological psychology*. New York: McGraw-Hill.

Likierman, A. (2009). The five traps of performance measurement. *Harvard Business Review*, *87*(10), 96–101.

Lim, C. (2018). From data to value: A nine-factor framework for data-based value creation in information-intensive services. *International Journal of Information Management*, *39*, 121–135.

Lipman, V. (2018, January 29). *Why Employee development is important, neglected and can cost you talent.* Forbes.com. Retrieved from https://www. forbes.com/sites/victorlipman/2013/01/29/why-development-planning-is-important-neglected-and-can-cost-you-young-talent/#7513ed86f633

Louisot, J., & Ketcham, C. H. (2014). *ERM enterprise risk management: Issues and cases.* Chichester: Wiley.

Luo, J. S., Hilty, D. M., Worley, L. L., & Yager, J. (2006). Considerations in change management related to technology. *Academic Psychiatry, 30*(6), 465–469. doi:10.1176/appi.ap.30.6.465

Mahoney, J. T., & Kor, Y. Y. (2015). Advancing the human capital perspective on value creation by joining capabilities and governance approaches. *Academy of Management Perspectives, 29*(3), 296–308. doi:10.5465/amp.2014.0151

Managing change: How law firms are answering the wake-up call, (July/ Aug 2009) *Law Practice, 35*(5), 32. Retrieved at https://www. americanbar.org/publications/law_practice_home/law_practice_archive/ lpm_magazine_articles_v35_is5_pg32.html

Marr, B. (2016, November 17). *Big data at Tesco: Real time analytics at the UK grocery retail giant.* Forbes.com. Retrieved from https://www.forbes.com/sites/bernardmarr/2016/11/17/big-data-at-tesco-real-time-analytics-at-the-uk-grocery-retail-giant/3/ #1d6afed51333

Marshall, A. (2017, November 17). Will Tesla's automated truck kill trucking jobs? *Wired.* Retrieved from https://www.wired.com/story/what-does-teslas-truck-mean-for-truckers/

Martin, B. L. (1995). The end of delegation? Information technology and the CEO. *Harvard Business Review, 73*(5), 161–172.

Mastrangelo, P. M., Prochaska, J., & Prochaska, J. (2008). How people change: The transtheoretical model of behavior change. *Master's Tutorial at the 23rd Annual Conference of the Society for Industrial and Organizational Psychology.* San Francisco, CA.

McDonald's. (n.d.). *McDonald's burgers: Hamburgers & cheeseburgers.* Retrieved from https://www.mcdonalds.com/us/en-us/full-menu/burgers.html

McFarlan, F. W., & Nolan, R. L. (2003, August 25). Why IT does matter. *Business School*. Retrieved from https://hbswk.hbs.edu/item/why-it-does-matter

Meffert, J., & Swaminathan, A. (2017, October). Management's next frontier: Making the most of the ecosystem economy. *McKinsey Insights*. Retrieved from https://www.mckinsey.com/business-functions/digital-mckinsey/our-insights/managements-next-frontier

Merzenich, M. M. (2013). *Soft-wired: How the new science of brain plasticity can change your life*. San Francisco: Parnassus.

Messina, C. (2017, May 30). The UN Leadership Framework: A catalyst for culture change at the UN. *UN Staff System College Blog*, Retrieved from http://www.unssc.org/news-and-insights/blog/un-leadership-framework-catalyst-culture-change-un/

Michie, L. et al. (2018). From her story, to our story. *Proceedings of the 2018 CHI Conference on Human Factors in Computing Systems - CHI 18*. https://doi.org/10.1145/3173574.31739

Miller, A. P., Lewis, S., & Waites, T. (2018). Essentials of advocacy. *Journal of the American College of Cardiology*, 71(22), 2598–2600.

Mind Content Tools Team. (n.d.). *Porter's five forces: Understanding competitive forces to maximize profitability*. Retrieved from https://www.mindtools.com/pages/article/newTMC_08.htm

Mingardon, S. et al., (2017, August 09). Digital-era change runs on people power. *BCG: The New New Way of Working Series*. Retrieved from https://www.bcg.com/encl/publications/2017/change-management-organization-digital-era-change-runs-people-power.aspx

Mintzberg, H. (1994). The fall and rise of strategic planning. *Harvard Business Review*, 72(1), 107–114.

Miriovsky, B. J., Shulman, L. N., & Abernethy, A. P. (2012). Importance of health information technology, electronic health records, and continuously aggregating data to comparative effectiveness research and learning health care. *Journal of Clinical Oncology*, 30(34), 4243–4248. doi:10.1200/jco.2012.42.8011

Mitra, A., Oregan, N., & Sarpong, D. (2018). Cloud resource adaptation: A resource-based perspective on value creation for corporate growth. *Technological Forecasting and Social Change, 130*, 28–38. doi:10.1016/j. techfore.2017.08.012

Morakanyane, R., Grace, A., & O'Reilly, P. (2017). Conceptualizing digital transformation in business organizations: A systematic review of literature. *BLED 2017 Proceedings*. 21.

Morris, P. W. (2011, February). *Brief history of project management.* Oxford: Oxford Univ. Press. Retrieved from http://www. oxfordhandbooks.com/view/10.1093/oxfordhb/9780199563142.001. 0001/oxfordhb-9780199563142-e-2

Morrison, C. (2016) Digital transformation strategy. *Digital Health & Care Institute, Glasgow*, Retrieved at: https://strathprints.strath.ac.uk/ 64342/

Murphy, W. (2017, August 09). *The nomenclature of artificial intelligence.* Retrieved from https://www.intuitiveaccountant.com/people-and-business/the-nomenclature-of-artificial-intelligence/#.W0ecD9JKhPY

Myrick, J. G. (2015). Emotion regulation, procrastination, and watching cat videos online: Who watches internet cats, why, and to what effect? *Computers in Human Behavior, 52*, 168–176. doi:10.1016/j. chb.2015.06.001

Nadler, D., Shaw, R. B., & Walton, A. E. (1995). *Discontinuous change: Leading organizational transformation.* San Francisco: Jossey-Bass.

National Research Council. (1999). *Being fluent with information technology.* Washington, DC: The National Academies Press. https://doi. org/10.17226/6482

Nelson, L. J. (2016, April 14). Uber and Lyft have devastated L.A.'s taxi industry, city records show. *LA Times*, Retrieved from http://www. latimes.com/local/lanow/la-me-ln-uber-lyft-taxis-la-20160413-story.html

Net Market Share. (n.d.). *Search engine market share.* Retrieved from https://www.netmarketshare.com/search-engine-market-share.aspx

Netburn, D. (2011, December 20). Talking twin babies, Nyan cat among youtube's top videos of 2011. *LA Times Blog*, Retrieved from http://latimesblogs.latimes.com/technology/2011/12/talking-twin-babies-nyan-cat-and-friday-top-youtubes-most-watched-videos-of-2011.html

Ng, S., Holm, E., & Ante, S. E. (2011, November 15). Buffett Bets $10. 7 Billion in Biggest Tech Foray. *Wall St. Journal*, Retrieved from https://www.wsj.com/articles/SB10001424052970204323904577037742077676990

Nicas, J. (2018, April 01). They tried to Boycott Facebook, Apple and Google. they failed. *NY Times*, Retrieved from https://www.nytimes.com/2018/04/01/business/boycott-facebook-apple-google-failed.html

Nilsson, N. J. (2010). *The quest for artificial intelligence: A history of ideas and achievements*. Cambridge: Cambridge University Press.

Niven, P. R. (2010). *Balanced scorecard step-by-step: Maximizing performance and maintaining results*. Hoboken: John Wiley & Sons.

Nocera, J. (2018, May 18). *Wells Fargo has shown us its contemptible values*. Bloomberg.com. Retrieved from https://www.bloomberg.com/view/articles/2018-05-18/wells-fargo-has-shown-its-customers-its-true-values-joe-nocera

Noria Corporation. (2007, August 09). *Middle managers are biggest obstacle to lean enterprise*. Retrieved from https://www.reliableplant.com/Read/7751/middle-managers-lean

O'Connor, G. C., Corbett, A. C., & Peters, L. S. (2018). *Beyond the champion: Institutionalizing innovation through people*. Stanford, CA: Stanford Business Books, an imprint of Stanford University Press.

O'Neil, G. M. (2017, February 22). Hiring algorithms are not neutral. *Harvard Business Review*, Retrieved from https://hbr.org/2016/12/hiring-algorithms-are-not-neutral

Oracle. Oracle Big Data. (n.d.). *Oracle*. Retrieved from https://www.oracle.com/big-data/guide/what-is-big-data.html

Ostherr, K. (2018, April 18). Facebook knows a ton about your health. Now they want to make money off it. *Washington Post*. Retrieved from https://www.washingtonpost.com/news/posteverything/wp/2018/04/18/facebook-knows-a-ton-about-your-health-now-they-want-to-make-money-off-it/

Ostrom, V. (1997). *The meaning of democracy and the vulnerabilities of democracies: A response to Tocqueville's challenge*. Ann Arbor: Univ. of Michigan Press.

Panorama. (2018, May 02). *The case for and against ERP customization*. Retrieved from https://www.panorama-consulting.com/the-case-for-and-against-erp-customization/

Panorama. (n.d.) *Why a business case is key to your digital transformation | digital transformation*. Retrieved from https://www.panorama-consulting.com/why-a-business-case-is-key-to-your-digital-transformation/

Pastin, M. (2018, January 21). The surprise ethics lesson of Wells Fargo. *Huffington Post*. Retrieved from https://www.huffingtonpost.com/mark-pastin/the-surprise-ethics-lesson_b_14041918.html

Perry, D. (2014, November 20). *Sex and Uber's 'Rides of Glory': The company tracks your one-night stands – and much more*. OrgeonLive.com. Retrieved from http://www.oregonlive.com/today/index.ssf/2014/11/sex_the_single_girl_and_ubers.html

Drucker, P. (n.d.). *Wikipedia Encyclopedia*. Retrieved at https://en.wikipedia.org/wiki/Peter_Drucker

Peterson, H. (2017, June 23). McDonald's shoots down fears it is planning to replace cashiers with kiosks. *Business Insider*. Retrieved from http://www.businessinsider.com/what-self-serve-kiosks-at-mcdonalds-mean-for-cashiers-2017-6

Pharoah, M. (2018, April 9). *Transforming change management with artificial intelligence (AI)*. Retrieved from https://www.andchange.com/transforming-change-management-artificial-intelligence-ai/

Polanyi, K. (1945). *Origins of our time: The great transformation*. London: V. Gollancz.

Prosci. (n.d.). *A change management office primer*. Retrieved from https://www.prosci.com/change-management/thought-leadership-library/a-change-management-office-primer

Pullen, J. P. (2014, November 04). Everything you need to know about Uber. *Time*. Retrieved from http://time.com/3556741/uber/

Regalado, A. (2013, December 30). Who coined 'cloud computing'? *Technology Review*. Retrieved from https://www.technologyreview.com/s/425970/who-coined-cloud-computing/

Reuters Graphics. (n.d.). *COBOL blues*. Retrieved from http://fingfx.thomsonreuters.com/gfx/rngs/USA-BANKS-COBOL/010040KH18J/index.html

Rezek, M. (2017, July 10). *The 3 types of fear that are hindering your growth as a leader*. Inc.com. Retrieved from https://www.inc.com/mary-rezek/overcome-3-main-reasons-people-fear-speaking-up.html

Rick, T. (2018, April 30). *Culture change is key in digital transformation*. Retrieved from https://www.torbenrick.eu/blog/culture/culture-change-is-key-in-digital-transformation

Roberto, D. A. (2005, February). Change through persuasion. *Harvard Business Review*. Retrieved from https://hbr.org/2005/02/change-through-persuasion

Robertson, P. (2007). *Always change a winning team*. London: Cyan.

Rogers, E. M. (2003). *Diffusion of innovations* (5th ed.). New York, NY: Free Press.

Rogow, B. J. (n.d.). *Demand chains*. Retrieved from http://demandchains.com/about-us/bruce-j-rogow

Rosenberg, M., Confessore, N., & Cadwalladr, C. (2018, March 17). *How Trump consultants exploited the Facebook data of millions*. Retrieved from https://www.nytimes.com/2018/03/17/us/politics/cambridge-analytica-trump-campaign.html

Rosenblat, A. (2016, April 07). The truth about how Uber's app manages drivers. *Harvard Business Review*, Retrieved from https://hbr.org/2016/04/the-truth-about-how-ubers-app-manages-drivers

Rushe, D. (2018, January 31). Facebook posts $4.3bn profit as Zuckerberg laments 'hard year'. *The Guardian*, Retrieved from https://www.theguardian.com/technology/2018/jan/31/facebook-profit-mark-zuckerberg

Sandberg, S. (2018, March 21) *Facebook post at 12:40 pm*. Retrieved from https://www.facebook.com/sheryl/posts/10160055807270177

Satell, G. (2014, September 21). *A look back at why blockbuster really failed and why it didn't have to.* Forbes.com. Retrieved from https://www.forbes.com/sites/gregsatell/2014/09/05/a-look-back-at-why-blockbuster-really-failed-and-why-it-didnt-have-to/#50ceb9431d64

Scania Connected Services. (2017, September 29). Scania One – the digital platform for connected services. *Scania.* Retrieved from https://www.scania.com/group/en/scania-one-the-digital-platform-for-connected-services/

Schneider, N. (2018, March 28). *Mark Zuckerberg: Give up Facebook control.* Retrieved from https://www.corpgov.net/2018/03/mark-zuckerberg-give-up-facebook-control/

Scriven, M., & Paul, R. (n.d.). *Defining critical th*inking. Retrieved from http://www.criticalthinking.org/pages/defining-critical-thinking/766

Segran, E. (2018, June 05). Luxury shoe startup Tamara Mellon just snagged $24 million. *Fast Company,* Retrieved from https://www.fastcompany.com/40581360/luxury-shoe-startup-tamara-mellon-just-snagged-24-million

Senge, P. M. (1990). *The fifth discipline: The art and practice of the learning organization.* New York: Doubleday.

Shah, N., Irani, Z., & Sharif, A. M. (2017). Big data in an HR context: Exploring organizational change readiness, employee attitudes and behaviors. *Journal of Business Research, 70,* 366–378. doi:10.1016/j.jbusres.2016.08.010

Siddiqui, F. (2018, March 31). Why D. C. is targeting the ride-hail industry. *Washington Post.* Retrieved from https://www.washingtonpost.com/local/trafficandcommuting/why-dc-is-targeting-the-ride-hail-industry/2018/03/31/ef01fca8-3473-11e8-94fa32d48460b955story.html

Sinclair, J., & Wilken, R. (2009). Sleeping with the Enemy: Disintermediation in Internet Advertising. *Media International Australia, 132*(1), 93–104. doi:10.1177/1329878x0913200110

Society for Risk Management. (2005). "RIMS Annual Report 2005." Retrieved from https://www.rims.org/aboutRIMS/AnnualReports/Documents/2005annualreport.pdf

Stackpole, B. (2008, March 13). *Five mistakes IT groups make when training end-users.* CIO.com. Retrieved from https://www.cio.com/article/ 2436969/training/five-mistakes-it-groups-make-when-training-end-users. html

Stollman, J. (2017, March 23). *Disruption: The new frontier for governance and risk professionals.* Retrieved from https://www. governanceinstitute.com.au/news-media/blog/2017/mar/disruption-the-new-frontier-for-governance-and-risk-professionals/

Streitfeld, D. (2018, January 19). Data-crunching is coming to help your boss manage your time. *NY Times,* Retrieved from https://www.nytimes. com/2015/08/18/technology/data-crunching-is-coming-to-help-your-boss-manage-your-time.html

Subbiah, K., & Buono, A. F. (2013). Internal consultants as change agents: Roles, responsibilities and organizational change capacity. *Academy of Management Proceedings, 2013*(1), 10721. doi:10.5465/ ambpp.2013.10721abstract

Swaminathan, A., & Meffert, J. (2017). *Digital @ scale: The playbook you need to transform your company.* Hoboken, NJ: John Wiley and Sons.

Taylor, B. (2016, November 28). How Domino's pizza reinvented itself. *NY Times.* Retrieved from https://hbr.org/2016/11/how-dominos-pizza-reinvented-itself

Tesco PLC. (n.d.). *Core purpose and values.* Retrieved from https://www. tescoplc.com/about-us/core-purpose-and-values/

Tesco: A measurable marketing case study. (2012, July 25). SmartCompany.com. Retrieved from https://www.smartcompany.com.au/ people-human-resources/managing/tesco-a-measurable-marketing-case-study

The Last Kodak moment?. (2012, January 14). *The economist.* Retrieved from https://www.economist.com/node/21542796

The Points Guy. (2014, December 22). *Insider series: What Uber drivers know about passengers.* Retrieved from https:// thepointsguy.com/2014/12/insider-series-what-uber-drivers-know-about-passengers/

The Role of the Sponsor in Bringing Change to a Project. (2011, July 06). *Bright hub project management*. Retrieved from https://www. brighthubpm.com/change-management/39144-sponsoring-a-change-management-initiative/

Thompson, D. (2018, May 17). Disneyflix is coming. And netflix should be scared. *The Atlantic*, Retrieved from https://www.theatlantic.com/ magazine/archive/2018/05/disneyflix-netflix/556895/

Tirone, J. (2018, February 16). Banks replaced with blockchain at international food program. *Bloomberg*, Retrieved from https://www. bloomberg.com/news/articles/2018-02-16/banks-replaced-with-blockchain-at-international-food-program

Ton, J. (2018, March 28*). It's all Greek to me – how executives can learn the language of technology*. Forbes.com. Retrieved from https://www. forbes.com/sites/forbestechcouncil/2018/03/28/its-all-greek-to-me-how-executives-can-learn-the-language-of-technology/

Tower, B. (2017*). How IoT Data collection and aggregation with local event processing work*. Retrieved from https://blog.equinix.com/blog/ 2017/11/29/how-iot-data-collection-and-aggregation-with-local-event-processing-work/

Toyota. (2014, December 09). *Nemawashi - Toyota production system guide*. Retrieved from http://blog.toyota.co.uk/nemawashi-toyota-production-system

Tozzi, J. (2007, July 13). Bloggers bring in the big bucks. *Business Week*, Retrieved from https://web.archive.org/web/20080215230339/http://www. businessweek.com/smallbiz/content/jul2007/sb20070713_202390.htm

Trello Enterprise. (n.d.). *Trello*. Retrieved from https://trello.com/ enterprise

Trust, T. (2017). 2017 ISTE Standards for Educators: From Teaching With Technology to Using Technology to Empower Learners. *Journal of Digital Learning in Teacher Education, 34*(1), 1–3. doi:10.1080/ 21532974.2017.1398980

Tufte, E. (2003, September 01). PowerPoint is evil. *Wired*. Retrieved from https://www.wired.com/2003/09/ppt2/

Twitter. (n.d.). *The Twitter rules*. Retrieved from https://help.twitter.com/en/rules-and-policies/twitter-rules

SARB Media Relations. (2018, June 06). Typical daily SA interbank settlements done in under 2hrs. *BIZ News*. Retrieved from https://www.biznews.com/global-investing/2018/06/06/sarb-blockchain-pilot-daily-interbank-settlements/

Uber Crunches User Data to Determine Where The Most 'One-Night Stands' Come From. (2014, November 18). *CBS news*. Retrieved from https://sanfrancisco.cbslocal.com/2014/11/18/uber-crunches-user-data-to-determine-where-the-most-one-night-stands-come-from/

UN Blockchain. (2018, April 04). *Multi-UN agency platform*. Retrieved from https://un-blockchain.org/category/wfp/

United Nations. (n.d.). *Laboratory for organizational change and knowledge (UNLOCK)*. Retrieved from https://www.unssc.org/featured-themes/united-nations-laboratory-organizational-change-and-knowledge-unlock/

University of Minnesota, perfect competition: A model. Retrieved from https://open.lib.umn.edu/principleseconomics/chapter/9-1-perfect-competition-a-model/

University of Minnesota, The nature of monopoly. Retrieved from https://open.lib.umn.edu/principleseconomics/chapter/10-1-the-nature-of-monopoly/

Valente, T. W., & Pumpuang, P. (2006). Identifying opinion leaders to promote behavior change. *Health Education & Behavior*, *34*(6), 881–896. doi:10.1177/1090198106297855

Van der Zee, I. (2002). *Measuring the value of information technology*. Hershey, PA: IRM Press.

Virgin Atlantic. (n.d.). *Our culture | virgin atlantic careers*. Retrieved from https://careersuk.virgin-atlantic.com/life-at-virgin-atlantic/culture

Virgin Group. (2016, February 17). *Our purpose*. Retrieved from https://www.virgin.com/virgingroup/content/our-purpose-0

Wagner, K. (2018, March 22). *Mark Zuckerberg says he's 'fundamentally uncomfortable' making content decisions for Facebook*. Recode.net.

Retrieved from https://www.recode.net/2018/3/22/17150772/mark-zuckerberg-facebook-content-policy-guidelines-hate-free-speech

Wakabayashi, D., & Shane, S. (2018, June 01). Google will not renew pentagon contract that upset employees. *NY Times*, Retrieved from https://www.nytimes.com/2018/06/01/technology/google-pentagon-project-maven.html

Wang, L. (2015). When the customer is king: Employment discrimination as customer service. *Virginia Journal of Social Policy and the Law*, 23, 249.

Wang, Y., Kung, L., & Byrd, T. A. (2018). Big data analytics: Understanding its capabilities and potential benefits for healthcare organizations. *Technological Forecasting and Social Change*, 126, 3–13. doi:10.1016/j.techfore.2015.12.01

Waychal, P. (2016). A framework for developing innovation competencies. *2016 ASEE Annual Conference & Exposition Proceedings*. https://doi.org/10.18260/p. 26321

Weinstein, P. V. (2014, November 05). To close a deal, find a champion. *Harvard Business Review*, Retrieved from https://hbr.org/2014/09/to-close-a-deal-find-a-champion

West, D. M. (2018). *The future of work: Robots, AI, and automation.* Washington, DC: Brookings Institution Press.

Westerman, G., Bonnet, D., & McAfee, A. (2014). *Leading digital: Turning technology into business transformation.* Massachusetts, MA: Harvard Business Review Press.

What is internal audit?. (n.d.). Retrieved from https://www.iia.org.uk/about-us/what-is-internal-audit/

Winsor, S. (2015, April 16). *Adopt big data, or else. Corn and soybean digest.* Retrieved from http://www.cornandsoybeandigest.com/precision-ag/adopt-big-data-or-else

Wong, J. C. (2018, March 22). Mark Zuckerberg apologises for Facebook's 'mistakes' over Cambridge analytica. *The Guardian*, Retrieved from https://www.theguardian.com/technology/2018/mar/21/mark-zuckerberg-response-facebook-cambridge-analytica

Workday. (2017) *Workday for financial services*. Retrieved from: https://
www.workday.com/content/dam/web/en-us/documents/datasheets/
datasheet-workday-for-financial-services-us.pdf

World Food Programme. (n.d.). *Building blocks*. Retrieved from http://
innovation.wfp.org/project/building-blocks

Wu, F., & Cavusgil, S. T. (2006). Organizational learning, commitment,
and joint value creation in interfirm relationships. *Journal of Business
Research, 59*(1), 81–89. doi:10.1016/j.jbusres.2005.03.005

Yang, C., Huang, Q., Li, Z., Liu, K., & Hu, F. (2017). Big data and cloud
computing: innovation opportunities and challenges. *International Journal
of Digital Earth, 10*(1), 13–53. doi:10.1080/17538947.2016.1239771

Yohn, D. L. (2017, July 25). Walmart won't stay on top if its strategy is
"Copy Amazon." *Harvard Business Review*. Retrieved from https://hbr.
org/2017/03/walmart-wont-stay-on-top-if-its-strategy-is-copy-amazon

Zabecki, D. (2015, May). *Military developments of world war I*.
Retrieved from https://encyclopedia.1914-1918-online.net/article/military_
developments_of_world_war_i

Zuckerberg, M. (2018, March 21) *Facebook post at 12:36 pm*, Retrieved
from https://www.facebook.com/sheryl/posts/1016ro0055807270177

INDEX